2008 Annual Meeting of the American Electrophoresis Society

Topical Conference at the 2008 AIChE Annual Meeting

Philadelphia, Pennsylvania
16 - 21 November 2008

ISBN: 978-1-61567-216-5

Printed from e-media with permission by:

Curran Associates, Inc.
57 Morehouse Lane
Red Hook, NY 12571

Printed by Curran Associates, Inc. (2009)

For permission requests, please contact AIChE
at the address below.

AIChE
3 Park Avenue
New York, NY 10016-5991

Phone: (203) 702-7660
Fax: (203) 775-5177

www.aiche.org

Additional copies of this publication are available from:

Curran Associates, Inc.
57 Morehouse Lane
Red Hook, NY 12571 USA
Phone: 845-758-0400
Fax: 845-758-2634
Email: curran@proceedings.com
Web: www.proceedings.com

TABLE OF CONTENTS

10a A Mixed Boundary Element Method/brownian Dynamics Approach for Simulating DNA Electrophoresis In Electrically Insulating Microfluidic and Nanofluidic Geometries

Jaeseol Cho and **Kevin D. Dorfman**. Department of Chemical Engineering and Materials Science, University of Minnesota, 151 Amundson Hall, 421 Washington Ave SE, Minneapolis, MN 55455

Virtually all microfluidic and nanofluidic devices are fabricated using electrically insulating materials (silicon, glass, or plastics), and their relatively complex features can lead to strongly inhomogeneous electric fields. We will present an approach for simulating DNA electrophoresis in these devices, based on a mixed Boundary Element Method/Brownian Dynamics (BEM/BD) approach. Our method is straightforward and computationally efficient, opening up a new route for simulating DNA electrophoresis in the vast array of devices currently being explored in the microfluidics and nanofluidics community. Moreover, electric field calculations by boundary elements are ideally suited for considering active nanodevices with dynamically changing shapes and electric fields. To demonstrate the accuracy of our method, we studied the collision of a large DNA molecule with an isolated, finite-sized, insulating post. As the electric field distribution for such a geometry is known, it provides an ideal test bed for our BEM/BD method. The BEM/BD simulation is indistinguishable from an equivalent BD simulation that uses the exact solution for the electric field. We will also present an analysis of various approaches aimed at reducing the computational cost of the simulations.

10b Effects of Gel Morphology on Predicting Optimal Times of Separation In Nanocomposite Gels

Jennifer Anne Pascal[1], Holly A. Stretz[2], Mario Oyanader[3], and Pedro E. Arce[3]. (1) Chemical Engineering, Tennessee Tech University, PH-214, Cookeville, TN 38505, (2) Department of Chemical Engineering, Tennessee Technological University, Cookeville, TN 38501, (3) Chemical Engineering, Tennessee Technological University, Department of Chemical Engineering, P.O. Box 5013, Cookeville, TN 38505

Research involving polymer gels with embedded nanoparticles of varying properties is quite attractive because of the multitude of potential applications, including separation of biomacromolecules, tissue scaffold growth, as well as the analysis of the efficiency and control of drug delivery. The presence of nanoparticles within gels has the potential to modify not only the gel morphology but also the electrokinetic properties of the gels; therefore, these nanoparticles may influence both the electrophoretic transport as well as electro-osmotic flows.

This project focuses on idealized capillary models whose characteristics are useful domains to mimic the gel morphology in the aforementioned nanocomposite gels. Such domains allow for the use of a nonuniform cross section and electrostatic potential along the capillary walls in order to capture the electrophoretic and electrostatic behaviors between the nanoparticles and the gel. Additionally, this communication will report details and illustrations of the modeling of the electroosmotic transport within the gels, and on the effect of various system parameters such as the convergent-divergent channel cross sections, electrostatic potential, and aspect ratio, among others. The research has effectively used the electrokinetic hydrodynamics (EKHD) concepts as introduced by Arce and Oyanader ([1], unpublished, [5]) coupled with the spatial averaging approach originally introduced by ([2],[3],[4]).

[1] Arce, P. and Oyanader, M. "Electrokinetic Hydrodynamics: An Introductory Graduate level Course." To be submitted to Chemical Engineering Education.

[2] Slattery, J. C., Momentum, Energy and Mass Transfer in Continua, Krieger, New York, 1981.

[3] Whitaker, S., Chem. Eng. Sci. 1985, 40, 1387.

[4] Cwirko, E. H.; Carbonell, R. G., . J. Colloid Interface Sci. 1989, 129, 513.

[5] Pascal, Jennifer, Arce, Pedro and Oyanader, Mario. "Electrokinetic-Hydrodynamics (EKHD): An Efficient Framework for Systematic Research." AIChE Poster Presentation, 2008.

10c Electrical Conductance of Aqueous Solutions of Surfactants with Added Electrolytes

Ezinwa O. Elele and Boris Khusid. Chemical Engineering, New Jersey Institute of Technology, University Heights, Newark, NJ 07102

To investigate the interaction of surfactants with ions, we measured the conductance of aqueous solutions of sodium dodecyl sulphate (SDS) and lithium dodecyl sulfate (LDS) with added electrolytes NaCl, KCl, and LiCl as a function of the surfactant concentration and the electrolyte concentration. Measurements were carried out on a broadband dielectric spectrometer BDS-80, Novocontrol. Experiments cover a broad range of the surfactant and electrolyte concentrations, from above the critical micelle concentration (CMC) to infinite dilution. Experimental data are compared with predictions for the concentration dependence of electrical conductance provided by theories for strong electrolytes. We consider the classical Onsager-Fuoss equation for a binary electrolyte, the Onsager-Kim equation for a multicomponent electrolyte, and equations for a three-component electrolyte recently derived within the mean spherical approximation. Deviation of theoretical predictions for strong electrolytes from experimental data on surfactant solutions provides insight into the formation of micellar aggregates below the CMC point.

10d Structure and Conformation of Protein-Surfactant Complexes In Gel Electrophoresis

Danilo C. Pozzo and Monica Ospinal. Chemical Engineering, University of Washington, 357 Benson Hall, Box 351750, Seattle, WA 351750

Sodium dodecyl sulfate polyacrylamide electrophoresis (SDS-PAGE) is one of the most widely used methods to carry out analytical separations of small poly-peptides and proteins. Despite its recognized importance, our understanding of the transport of surfactant denatured proteins during electrophoresis is based greatly on assumptions of protein conformation inside the gel and on macroscopic measurements of their mobility. Contrast-variation SANS is used to provide, for the first time, a structural picture of the conformation of protein-surfactant complexes during this separation. Our recent SANS observations are discussed and related to transport models (Ogston model and reptation model) that have been proposed to occur during SDS-PAGE. The SANS technique is also applied to the analysis of alternative anionic surfactants to optimize separation efficiency based on the fundamental nano-structure of the complex.

10e Optimal Separation Times In An Electrical Field Flow Fractionation (EFFF) Separator: Effect of Electroosmotic Flow

Jennifer Anne Pascal[1], Mario Oyanader[2], and Pedro E. Arce[2]. (1) Chemical Engineering, Tennessee Tech University, PH-214, Cookeville, TN 38505, (2) Chemical Engineering, Tennessee Technological University, Department of Chemical Engineering, P.O. Box 5013, Cookeville, TN 38505

Separation of biomacromolecules for pharmaceutical processes is a multi-million dollar effort where even the smallest increase in the separation efficiency has a very healthy economical impact on the process or application. Antibiotics, proteins, and DNA are some key examples of biomacromolecules where new techniques and approaches for separation and or characterization are require. Electrokinetic-based methods have been used in a wide variety of applications not only related to separations but also to environmental remediation and drug delivery, to name a few. The difficulty of these methods is the efficient and consistent design of separation protocols due, in part, to the fact that the role of EOF on dispersion of the solute within the channel is not completely understood. These aspects become even more relevant when the dimensionsof the channel are small.

One of the EK-based applications is EFFF. In this separation technique, an orthogonal applied field (i.e. gravitational, electrical, etc.) drives the motion of the charged molecules towards the channel walls; thus, the solute that is most susceptible to the influence of the field locates closest to the walls, and, depending on the solute properties, different species reach the wall at different locations.

In this presentation, the authors will discuss a model that can be applied to an EFFF separation device to determine the effect of electro-osmotic flow on the separation efficiency. In particular, optimal times of separation will be computed and parametrically analyzed for a wide range of values. Principles of electrokinetic hydrodynamics (EKHD, see Arce et al [1]) allow for the connection to the convective-diffusive transport with the aid of the spatial averaging approach. Results of the research effort will predict the effects of electroosmosis on the optimal separation times to design and or optimize an EFFF separation device where electroosmotic flow could play a role.

[1] Arce, P.; M. Oyanader, and J. Pascal, "Electrokinetic-Hydrodynamics (EKHD): An Introductory Graduate level Course." To be submitted to Chemical Engineering Education.

10f Molecular Dynamics Simulations of the Effect of Dc and AC Electric Field on Polyelectrolyte

Boris Khusid, Chemical Engineering, New Jersey Institute of Technology, University Heights, Newark, NJ 07102, Tak S. Lo, Levich Institute, City College of CUNY, 140th Street & Convent Avenue, New York, NY 10031, and Joel Koplik, Physics and Levich Institute, City College of the City University of New York, 140th Street and Convent Avenue, New York, NY 10031.

Polyelectrolytes are molecules consisting of repeating ionizable groups which dissociate in water or other polar solvents, leaving a charged polymer backbone. A highly ionized polyelectrolyte in solution attracts small mobile counterions of opposite charge, partially screening the backbone charge, and mediating the intermolecular Coulomb repulsion. The difficulty in studying polyelectrolyte-counterions complexes resides in the delicate interplay of the long range electrostatic interactions, short range intermolecular interactions, and thermal energy, all of comparable magnitude. In contrast to previous studies, which considered equilibrium properties of polyelectrolyte-counterions complexes alone, we used molecular dynamics (MD) simulations to study the spatio-temporal dynamics of charge fluctuations around a flexible polyelectrolyte molecule at charge densities above and below the classic counterion condensation threshold. We report the results of MD simulations of the behavior of a polyelectrolyte subject to DC and AC fields. Theoretical predictions are compared with experimental data on electrophoretic mobility, polarization, and birefringence of dilute polyelectrolyte solutions.

INSULATOR-BASED DIELECTROPHORESIS OF PROTEIN PARTICLES USING DIRECT CURRENT ELECTRIC FIELDS

Sandra Ozuna-Chacón, Blanca H. Lapizco-Encinas and Marco Rito-Palomares
Tecnológico de Monterrey, Monterrey, NL, México

Abstract

Dielectrophoresis is a method of rapid response with sufficient selectivity for manipulation and separation of bioparticles, such as: microorganisms and biomolecules. Due to the great importance of proteins in the biotechnological and pharmaceutical processes, the present study demonstrates the potential of the insulator-based dielectrophoresis (iDEP) and DC electric fields to manipulate and concentrate protein solutions, using bovine serum albumin as a model. Samples containing fluorescently labeled bovine serum albumin protein were manipulated inside a microchannel made from glass that contained an array of cylindrical insulating structures. DC electric fields were applied and the dielectrophoretic response of the particles was observed. It was shown that the magnitude of the applied electric field and the conductivity and pH of the suspending medium have a strong effect on the dielectrophoretic response of the protein particles.

Introduction

Miniaturization has brought important advantages to bioseparation technology. Numerous fields, including environmental, pharmaceutical and biochemical, have beneficiated from the advances of microanalytical systems. There is a growing interest on the development of separation techniques that can applied in microscale. Dielectrophoresis (DEP) is a non-destructive electrokinetic transport mechanism, widely employed in microfluidics to manipulate bioparticles. DEP is the movement of particles in a nonuniform electric field, due to polarizations effects, and it can occur in alternating current (AC) or direct current (DC) electric fields. This technique was described for the first time by Pohl in 1951 [1]. The majority of the studies reported in literature, have been carried out employing electrodes, where non-uniform electric fields are generated using an array of electrodes and AC fields [2-3]. However, it is also possible to generate non-uniforms electric fields by means of electric insulator materials. Insulator-based DEP (iDEP) is an alternative to the traditional electrode-based-DEP. In this technique, only two electrodes are needed since the nonuniformity of the electric field is generated with insulating structures [4]. There are a number of advantages offered by iDEP: fabrication microdevices for iDEP are economical and simpler, leading to more robust devices that can conserve their functionality in despite of fouling.

The manipulation of protein molecules has been demonstrated with electrode-based DEP. Washizu *et al.* in 1994 [5], reported the dielectrophoretic manipulation of the proteins: avidin, concanavalin, chymotripsinogen, and ribonuclease A, employing a set of corrugated electrodes and AC electric fields. In 1998 Bakewell and collaborators [6] demonstrated the dielectrophoretic manipulation of the avidin, using polynomial electrodes. In 2004, Zheng and col. [7], reported the dielectrophoretic manipulation of protein of bovine serum albumin (BSA), using quadrupole electrodes. The manipulation of proteins molecules has not been demonstrated with iDEP.

The present study reports the manipulation and concentration of BSA employing insulator-based-DEP and DC electric fields. The protein BSA was concentrated inside a glass microchannel containing an array of cylindrical insulating posts. These results showed that the magnitude of the applied electric field and suspending medium properties (conductivity and pH) have a strong effect on the dielectrophoretic response of the protein particles. The results presented here are the first demonstration on manipulation and concentration of protein employing DC-iDEP.

Theory of dielectrophoresis

The magnitude of the dielectrophoretic force depends on the intensity of the applied electric field, particle size, and on the dielectric properties of particles and suspending medium [8, 9]. The dielectrophoretic force acting on a spherical particle is defined as:

$$F_{DEF} = 2\pi\varepsilon_0\varepsilon_m r^3 \operatorname{Re}\lfloor f_{CM} \rfloor \nabla(E \cdot E) \tag{1}$$

where ε_0 is the permittivity of free space, ε_m is the relative permittivity of suspending medium, r is the particle radio and $\operatorname{Re}\lfloor f_{CM} \rfloor$ it is the real part of the Clausius-Mossotti (CM) factor. According to Hughes [10] when low frequency electric field are used (frequency ≤ 100 kHz), the CM factor can be written as:

$$f_{CM} = \left[\frac{\sigma_p - \sigma_m}{\sigma_p + 2\sigma_m}\right] \tag{2}$$

From Eqns. 1 and 2 it is possible to observe that the dielectrophoretic force exerted on a particle can be positive or negative depending on the sign of the CM factor. Additionally, the dielectrophoretic force depends on the intensity of the electric field, the particle size and dielectric properties of the particle and suspending medium. These operating conditions can be varied to manipulate the DEP force order to achieve separation and concentration of particles. The dielectrophoretic force is the second order with respect to the applied electric fields (Eqn. 1); at low applied electric fields, low electric fields gradients are produced, leading to negligible dielectrophoretic force. Therefore, the applied electric field has to be high enough for DEP to become significant and immobilize particles. The two regimes of iDEP are *streaming* and *trapping* DEP. *Streaming DEP* occurs when the dielectrophoretic force overcomes diffusion but does not overcome the electrokinetic flow. *Trapping DEP* is the reversible immobilization of particles in dielectrophoretic traps and it occurs when the DEP force overcomes diffusion and electrokinetic flow (as well as pressure-driven flow if present) [13-15]. The dielectrophoretic velocity is of second order effect of the electric field, can be expressed as:

$$v_{DEP} = -\mu_{DEP}\nabla E^2 \tag{3}$$

The electrokinetic flow is proportional to the electric field (Eqn. 4). The electrokinetic flow comprises the effects of electroosmosis and electrophoresis on particle motion relative to a fixed channel. For a glass system the electrokinetic velocity is obtained employing the electrokinetic mobility as follows:

$$v_{EK} = \mu_{EK}E = (\mu_{EO} - \mu_{EP})E \tag{4}$$

where v_{EK} is the electrokinetic velocity, μ_{EK} the electrokinetic mobility and E is the applied electric field. This research work is focused on the demonstration of protein trapping employing iDEP. One of the objectives is to study the effect of bulk medium properties on the dielectrophoretic response of protein particles. The bulk or suspending medium properties have an effect on the zeta potential of the microchannel wall, which affects the electroosmotic mobility:

$$\mu_{EO} = \frac{\zeta \varepsilon_m}{\eta} \qquad (5)$$

where ζ is the zeta potential of the microchannel or capillary wall, ε_m and η are the permittivity and viscosity of the suspending medium respectively. In this work, suspending mediums with conductivities varying from 25 to 100 µS/cm were employed. The conductivity for BSA has been reported by Zheng *et al.* [7] as σ_p = 25 µS/cm. Employing Eqn. 2 it is possible to calculate the CM factor for the protein particles suspended in each one of the three different bulk medium employed, the values of the CM factors are reported in Table 1.

Table 1. Clausius-Mossotti factors of protein particles

Conductivity of the protein particles (µS/cm)	Conductivity of the suspending medium (µS/cm)	CM factor
25	25	0.00
25	50	-0.29
25	100	-0.33

Materials and Methods

Equipment description. A schematic representation of the experimental setup is shown in Figure 1. Experiments were conducted employing microchannel made from glass that was 10-µm deep, 10.16-mm long and 2-mm wide. An array of cylindrical insulating posts was etched inside the microchannel; the posts had a diameter of 440 µm, a height of 10 µm and a center-to-center separation of 520 µm. In order to prevent particles from crashing against the posts and plugging the system, "dove-tail" geometry was used for the first row of posts on either side (Fig. 1). A 3000 V high voltage sequencer, model HVS448 (LabSmith, Livermore CA) was used as power supply. Protein particles were dyed with fluorescein isothiocyanate (Sigma, Toluca, Mexico) following a standard protocol provided by the dye manufacturer. A sample of 40 µl of protein solution at a concentration of 30 mg/ml is added to the inlet reservoir of the microchannel, resulting in an approximate concentration of 15 mg/ml of protein in the microchannel. The dielectrophoretic behavior of protein particles was recorded in the form of videos by employing an inverted epifluorescence video microscope for microfluidics, model SVM340 (LabSmith, Livermore CA). A 4X microscope objective was used for all experiments. Prior to experimentation the microchannel was filled with an aqueous solution with adjusted pH and conductivity. The pH of this solution was adjusted to a value of 8 or 9 by adding NaOH and, the conductivity to values of: 25 µS/cm, 50 µS/cm and 100 µS/cm, by adding K_2HPO_4.

Results and Discussion

Figure 2 shows some of the results obtained using different pH and conductivity of the suspending medium. The effect of the magnitude of the applied field can be observed by comparing Figures 2a and 2b, where it is shown that by increasing the applied electric field to 900 and 1200 V/cm (employing a suspending medium pH=8 and σ=25 µS/cm), respectively, a higher dielectrophoretic force is obtained. This dielectrophoretic overcomes the electrokinetic force, producing negative dielectrophoretic trapping. From Table 1, when a suspending medium with a conductivity of 25 μS/cm is employed, it is not expected that the protein particles will exhibit a dielectrophoretic response, since under these conditions CM factor is zero. However, the results shown in Figures 2a and 2b, demonstrate that the protein particles exhibited negative DEP behavior. This can be explained as follows: the conductivity value reported for the protein particles of 25 μS/cm [7] is an approximate value obtained employing AC electric fields. Under the operating conditions used in our work, where DC electric fields were applied, it is possible for the protein particles to behave less conductive, which would lead to a negative value for the CM factor.

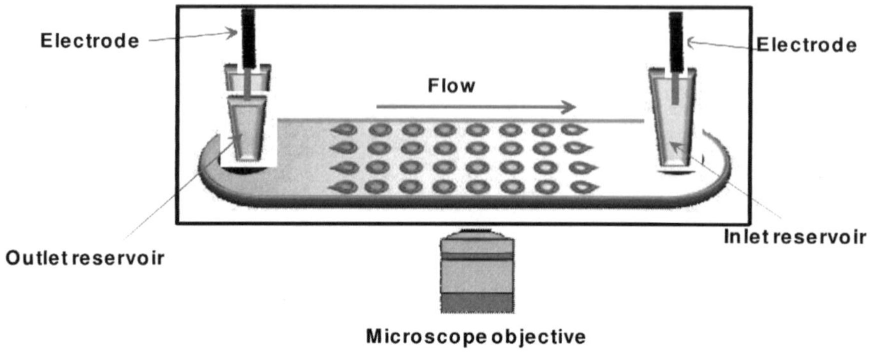

Figure 1. Experimental set-up showing the layout of a microchannel.

The effect of suspending medium conductivity can be observed by comparing Figures 2a and 2c. From the figures it can be seen that by increasing the conductivity of the suspending medium to 100 μS/cm (employing a pH of 8 and almost the same electric field), protein particles show stronger dielectrophoretic trapping, since protein bands appear brighter. These results show that by increasing the suspending medium conductivity, the magnitude of the negative CM factor increases (Eqn. 2 and Table 1), which in turn increases the dielectrophoretic force.

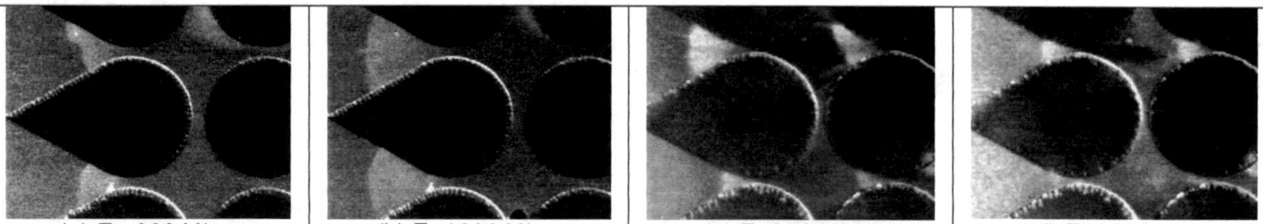

(a) E= 900 V/cm pH=8, σ=25 μS/cm	(b) E=1200 V/cm pH=8, σ=25 μS/cm	(c) E=950 V/cm pH=8, σ=100 μS/cm	(d) E=900 V/cm pH=9, σ=25 μS/cm

Figure 2. Dielectrophoretic response of protein of particles (shown green-yellow) inside a microchannel with cylindrical insulating structures, flow direction is from left to right, post diameter is 440 μm; a) negative dielectrophoretic trapping, b) strong dielectrophoretic trapping, c) strong dielectrophoretic trapping, d) Strong dielectrophoretic trapping.

The pH of suspending medium also affects the dielectrophoretic response, by comparing Figures 2a and 2d with a conductivity of 25 μS/cm and pH of 8 and 9, respectively, and the same electric field) it is possible to observe that by increasing the pH of the suspending medium, the dielectrophoretic trapping is lower.

A series of experiments were carried out to observe the effect of conductivity and pH suspending medium on the magnitude of the minimum electric field required to achieve dielectrophoretic trapping. Six different suspending mediums with conductivity values of 25, 50, and 100 μS/cm and pH values of 8 and 9 were employed. The experiments were performed by slowly ramping up the applied field and observing the dielectrophoretic response of the protein particles, in order to identify the lowest required field to achieve dielectrophoretic trapping. The results are presented in Figure 3.

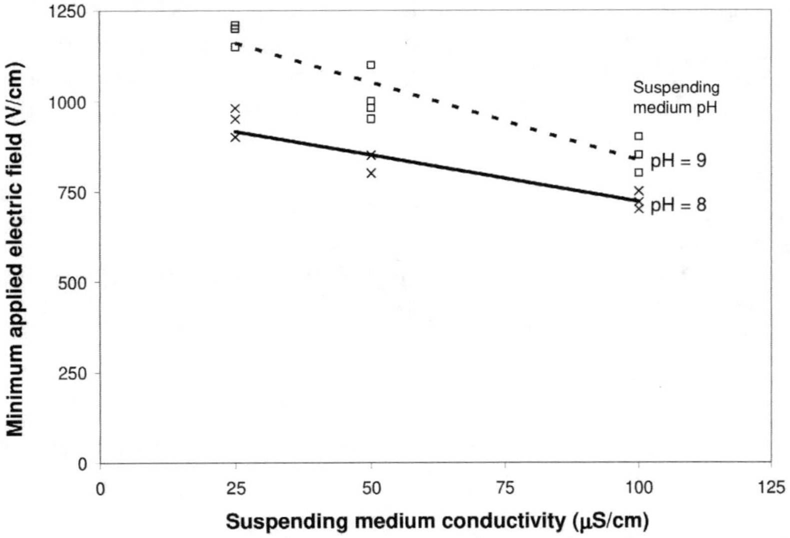

Figure 3. Minimum applied electric field required to achieve dielectrophoretic trapping of protein particles, results obtained by varying the conductivity and pH of the suspending medium.

Effect of suspending medium pH: As it can be observed in Figure 3, the minimum electric field necessary to achieve dielectrophoretic trapping increases with the pH of the suspending medium. These results can be explained as follows, a change in pH affects the zeta potential of the glass microchannels. The higher the pH of the suspending medium, the higher the desprotonization of the silanol groups on the glass surface, leading to a higher zeta potential that results in a stronger EOF (higher electroosmotic mobility Eqn. 5) [11, 12].

Effect of suspending medium conductivity: From Figure 3 it is possible to observe that increasing the conductivity of the suspending medium, decreases the magnitude of minimum electric required to achieve dielectrophoretic trapping. When the conductivity of the medium is increased, the magnitude of the negative CM factor also increases, leading to an increase on the magnitude of the dielectrophoretic force, improving dielectrophoretic trapping of protein particles. As shown in Table 1, the CM factor increases when the conductivity of the suspending medium increases.

These results demonstrate a way of manipulating dielectrophoretic separation of particles by varying the conductivity and pH of the suspending medium. The optimal operating conditions to reduce energy consumption are to use the lowest pH and highest conductivity possible for the suspending medium.

Conclusions

The present study reports the manipulation of protein particles employing insulator-based dielectrophoresis and DC electric fields. The results showed that the protein particles can be trapped with negative dielectrophoresis as a function of the magnitude of the applied electric field, and the pH and conductivity of the suspending medium. Higher applied electric fields produced strong negative dielectrophoretic trapping of BSA protein particles. Increasing the conductivity of the suspending medium, increases dielectrophoretic trapping, while increasing the pH of the suspending medium decreases dielectrophoretic trapping. The properties of the suspending medium are conditions that can be varied in order to manipulate dielectrophoretic separations of bioparticles. These results demonstrate the great potential of iDEP as technique for bioseparations.

Acknowledgements

The authors would like to acknowledge the financial support provided by CONACYT grant No. CONACYT-CB-2006-53603 and *Cátedras de Investigación* (CAT161 and CAT162) of *Tecnológico de Monterrey*.

References

1. Pohl H., (1951). "The motion and precipitation of suspensoids in divergent electric fields," *Applied Physics,* 22; pp.869-871.
2. Li H. and Bashir R., (2002). "Dielectrophoretic separation and manipulation of live and heat-treated cells of Listeria on microfabricated devices with interdigitated electrodes," *Sensors and Actuators B:Chemical*, 86 (2-3); pp.215-221.
3. Medoro G., Manaresi N., Leonardi A., Altomare L., Tartagni M. and Guerrieri R., (2002). "A lab-on-a-chip for cell detection and manipulation," in *IEEE Sensors, Proc. IEEE*, Orlando, FL, USA, pp. 472-477.
4. Wang X., Huang Y., Gascoyne P.R.C. and Becker F., (1997). "Dielectrophoretic manipulation of particles," *IEEE Transactions on Industry-Applications-Society*, Lake Buena Vista, FL, 33 (3); pp.660-669
5. Washizu M., Suzuki S., Kurosawa O., Nishizaka T. and Shinohara T., (1994). "Molecular dielectrophoresis of biopolymers," *IEEE Transactions on Industry Applications* 30(4); pp. 835-843.
6. Bakewell D., Hughes M., Milner J.J. and Morgan H., (1998). "Dielectrophoretic manipulation of avidin and DNA," in *20th International Conference of the IEEE in Medicine and Biology Society*, 20(2); pp.1079-1082.
7. Zheng L.J., Brody P. and Burke P., (2004). "Electronic manipulation of DNA, proteins, and nanoparticles for potential circuit assembly," *Biosensors & Bioelectronics*, 20; pp.606-619
8. Pohl H., (1978). *Dielectrophoresis.* Cambridge: Cambridge University Press, pp.573.
9. Halaka F.G., (2003). "Dielectrophoretic dynamic light-scattering (DDLS) spectroscopy", *Proceedings of the National Academy of Sciences of the United States of America,*

100(18); pp.10164–10169.

10. Hughes, M.P., Morgan, H., Rixon, F.J., (2001). “Dielectrophoretic manipulation and characterization of herpes simplex virus-1 Capsids,” *European Biophysics Journal* 30; pp. 268–272.

11. Kirby B.J. and E.F. Hasselbrink, (2004). “Zeta potential of microfluidic substrates: 1. Theory, experimental techniques, and effects on separations,” *Electrophoresis*, 25; pp.187-202.

12. Adamson A. and Gast A., (1997). *Physical Chemistry of Surfaces*. 6a ed. United States of America; pp.808.

13. Cummings E.B. and Singh A.K., (2000). “Dielectrophoretic trapping without embedded electrodes,” *SPIE: Conference on Microfluidic Devices and Systems III*, 4177; pp.164-173.

14. Cummings E.B. (2002). “A Comparison of Theoretical and Experimental Electrokinetic and Dielectrophoretic Flow Fields,” in *32nd AIAA Fluid Dynamics Conference and Exhibit* 2002. St. Louis, Missouri; paper 2002-3193, pp.1-17.

15. Cummings E. and Singh A., (2003). “Dielectrophoresis in Microchips Containing Arrays of Insulating Posts: Theoretical and Experimental Results,” *Analytical Chemistry*, 75; pp.4724-4731.

Dielectrophoretic Characterization of Red Blood Cells

Kaela Leonard, Eric Rutan, Sheena Reeves, Ashley Pate, Megan Walton, Sarah Thompson, Adrienne Minerick
Dave C. Swalm School of Chemical Engineering, Mississippi State University

Abstract

Medical microdevices have been touted as the doctor's 'toolkit of the future' for their potential ability to perform a variety of diagnostic tasks while utilizing only a single drop of blood. This work describes a unique lab-on-a-chip device that performs blood sample preparation via cell rupture to facilitate access to intracellular biomarkers. Traditional rupturing is accomplished via chemical lysis, which can leave unwanted chemicals in microdevice channels and interfere with subsequent separations, purifications, and analysis of desired proteins or other biomarkers. A recent alternative advanced by our lab group is rupturing the blood cells via dielectrophoresis; it is chemical free and entirely field dependent, so the intracellular contents of the red blood could be subsequently analyzed on the same chip. The purpose of this work is to quantify the rate at which red blood cells rupture when exposed to a 1kHz alternating current electric field. The two variables tested herein are the blood type and the age of the blood after donation has occurred. Rupturing behaviors at a 1 kHz dielectrophoretic field were quantified as a function of age for blood types A+, B+, AB+, O+ and O-. Age dependency was tested by conducting rupture experiments of the cells every three days after blood donation. The work presented herein focuses on the rupturing rate of five different blood types on the day of donation. Erythrocyte rupturing rates were quantified by examining the still frames of video of ruptured red blood cells to determine the rupture rate. Rupturing rate varied based on age and blood type from no rupturing to approximately 90% of the original number of red blood cells in 14 minutes. This work would advance and simplify medical microdevices by enabling one step lysis in line with separation, purification, and analysis for fully integrated diagnostic blood tests.

Introduction

Microdevices, also known as lab-on-a-chip devices or micro total analysis systems, emerged in the 1990's for the use of cell identification and analysis. These devices are seen as one of the key growth industries for the 21st century [1]. The accessibility of these devices make them very attractive for use as point-of-care devices in situations such as roadside emergencies, natural disasters, or for use in remote field locations where access to traditional bench scale equipment is difficult. These miniature devices require dedicated techniques for cell separation and subsequent cell analysis. One such technique explored herein is the use of dielectrophoretic electric fields for red blood cell hemolysis.

Dielectrophoresis (DEP) is the movement of particles in a non-uniform alternating current (AC) field. When a charged particle is subjected to a non-uniform AC field, the particles become polarized and move in accordance with their dielectrophoretic force. The particle will experience a force towards the high or low electric field density regions based on its polarizability [2]. If the particle is more polarizable than the medium the particle will move towards the region of high field density. If the particle is less polarizable than the medium, it will move towards the low field density regions. Bioparticles and cells are more complex than homogenous colloidal particles and therefore demonstrate more complex behaviors in dielectrophoretic fields [3]. Dielectrophoresis has been explored as a way to quickly separate and identify living cells [4]. Previous red blood cell dielectrophoretic studies have focused on frequencies nearing 1MHz and have recorded the movement of the red blood cells as function of age and blood type [3,5]. While these tools have focused on intact cell manipulations, our work has revealed what 1 kHz dielectrophoresis can rupture cells such that access to sub cellular analytes is possible.

Therefore dielectrophoresis may be useful for cellular preparation for subsequent proteomic and genomic analysis. Common methods of cell destruction include chemical lysis or electroporation. The most common chemicals used for red blood cell destruction are hypochlorous and hypobromous acid [6]. The use of such chemicals in microdevices can present complex design challenges. These challenges include overcoming problems with on chip storage of reagents, flowing and contacting the chemicals with the cells, mixing to insure adequate contact time, the addition of a washing step after separation and the possibility of unwanted effects in the final analysis [7]. Chemical lysis can be caustic and cause harmful or unwanted interactions in later analyses.

Through the fields of electroporation and pulsed field electrophoresis, it is known that cells can also be irreversibly damaged either accidentally or as a means of cell rupturing using electric fields much higher than 10kHz [8]. It has been demonstrated that the membranes of both free flowing cells and those adhered to the bottom of a channel can be permeabilized by electroporation using a mechanical valve to create direct current pulses. This technique eliminates the cost of microfabricated electrodes that are often found in cell lysis devices [9]. Cell destruction by dielectrophoresis however, uses much weaker fields than by electroporation and is usually due to a breakdown of the membrane surrounding the cell [8]. Lysis by dielectrophoresis has many advantages such as the lack of chemicals allowing for integration of lysis and analysis on a single microchip and the low power needed for a dielectrophoretic approach would enable device portablilty. One application of dilectrophoretic cell lysis is in medical diagnostics. Blood is a convenient diagnostic fluid because it is easy to attain uncontaminated samples and provides chemical information about systemic physiology. Blood is composed of erythrocytes (red blood cells = 45%), leukocytes (1%) and plasma (55%) [10]. Due to erythrocyte prevalence in blood, this work focuses on red blood cells.

The average adult has 4.2 million to 6.2 million red blood cells per microliter. For males the average is 5.5 million red blood cells per microliter as opposed to 4.5 million red blood cells per microliter for females [10]. Red blood cells are biconcave disks of diameter 7 microns and thickness 2 microns; this shape maximizes the surface through which the cells can transport oxygen and carbon dioxide. The life span of a red blood cell in circulation is approximately 120 days and undergoes physiological changes during that time such as increased membrane fragility which can make red blood cells more susceptible to hemolysis [10].

An incredibly important factor when working with blood is the blood type due the varying polysaccharide molecules, or complex carbohydrate antigens found on the cell membrane. The ABO blood typing system was discovered by Landsteiner in 1900 [11]. This blood typing system is defined by the presence or absence of two main antigens. Type A blood is classified by the presence of the A antigen, whereas type B blood is classified by the presence of the B antigen. Type AB blood has both A and B antigens and type O blood has neither antigen [10]. Another major factor in the ABO system is the presence or absence of powerful antibodies in the surrounding cell plasma. Normally the body does not produce antibodies unless exposed to the antigen; an exception is for the antibody against the antigen that the body does not naturally produce. Antigen A and antigen B only vary by a subsitution on the terminal end of the antigen. Antigen A has N-acetyl-D-galactosamine, whereas B has N-acetyl-D-glucosamine. However because the number of antigens per cell can be as high as 1.5 million this small change makes a large cellular difference [7]. *In vivo* hemolysis, the natural demise of red blood cells in the body, is largely controlled by a protein binding to the red blood cell membrane, specifically to the antibody that has bound to the surface antigen to form an antigen-antibody complex. This complement protein then binds and induces hemolysis [10]. Due to the *in vivo* role of antigens in red blood cell destruction it is hypothesized that blood type will influence *in vitro* red blood cell destruction. In addition to the A and B antigens there is a third antigen that is important in matching blood types during transfusions [10]. This antigen is known as the Rhesus factor. If the Rh antigen is present, the blood type is classified as positive and if it is absent the blood type is positive [12]. Therefore, the eight resulting blood types are A+,B+,O+,AB+,A-,B-,O-,AB-. Due to the fact that the antigens that define blood type are found on the cell membrane and the cell membrane plays a major role in the ability of the cell to rupture, it is hypothesized that the rupturing of the red blood cell will be dependent on the blood type based on previously presented initial results [13].

The purpose of this work is to quantify the rate at which red blood cells rupture when exposed to a 1kHz alternating current electric field. The two variables tested herein are the blood type and the age of the blood after donation has occurred. Blood type is an important factor because it is an easily known chemical variant in the population that can be used to assess antigen role in membrane stability. Also testing the same blood type from various donors can assess the reproducibility of the experimental technique. Age of the blood after donation is important because often field tests will need to be later verified in a laboratory setting. Also blood banks could use this tool (i.e. rupture times) as an indicator of the stability and health of the stored samples. Rupturing behaviors at a 1 kHz dielectrophoretic field were quantified as a function of age for blood types A+, B+, AB+, O+ and O-. The was to assess the impact of the A and B antigens as well as the presence or absence of the Rhesus factor. Age dependency was tested by conducting rupturing experiments of the cells approximately every three days after blood donation.

Materials and Methods

Blood Sample Preparation

Whole blood samples representing blood types A+, B+, AB+, O+ and O- were obtained via venipuncture from anonymous donors at the Longest Student health center. This blood was then drawn into a vacutainer with 1.8 mg K_2 EDTA anticoagulant per mL of blood. After donation the blood vial was stored at $5^{\circ}C$ in a Biosafety Refrigerator. Just prior to experiments, whole blood was mixed with an isotonically matched phosphate buffer saline (0.140molar NaCl, 0.02489 molar KH_2PO_4, 0.00907molar K_2HPO_4) before being injected in a 600:1 V/V mixture into a custom fabricated microdevice for rupture.

Microdevice Fabrication

Custom microdevices were fabricated using the soft photolithography technique developed by Whitesides [14]. A high resolution AutoCAD mask was printed at 32,512 dpi by Fineline Imaging (Colorado Springs, CO). The autocad design can be seen below in Figure 1. Silica wafers were washed with various organic solvents (ethanol, isopropanol and acetone) and surface functionalized with sodium hydroxide. After drying, negative photoresist was spin coated onto the wafer at an even thickness of 200 microns. After baking the photoresist the AutoCAD mask was placed over the coated wafer and exposed to a UV light to imprint the mask design on the photoresist layer. After the masked photoresist was baked, the wafer was placed in the photoresist developer such that the features of the mask remained on the wafer and any excess photoresist was washed off with isopropanol and acetone. Polydimethylsiloxane (PDMS) elastomer (Dow Corning) and a curing agent were then cast over the design on the wafer and baked overnight at 65°C. The PDMS casting was gently peeled off the wafer, and sample ports were punched out with a biopsy punch. After electrodes were embedded in the PDSM the device was treated in a UV Ozone Generator for 10 minutes and thus irreversibly bound to a glass slide.

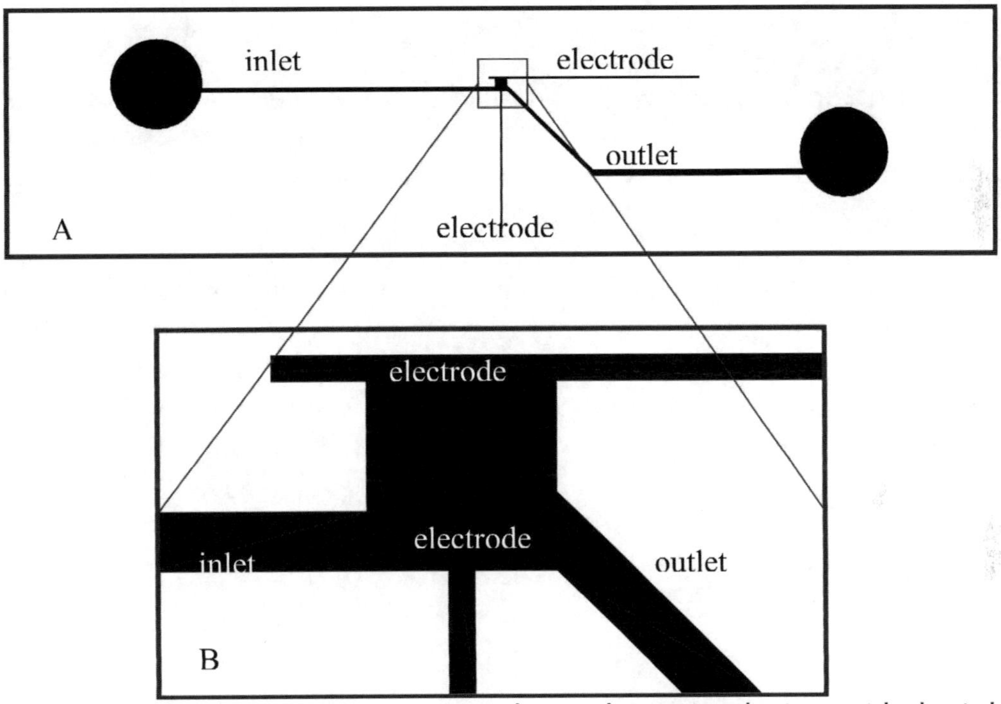

Figure 1. AutoCAD Mask of Device Design. A) Shows the entire device, with the inlet and outlet ports at the ends and the perpendicular lines for electrodes. B) Shows the experimental chamber of dimensions 800x800 microns with inlet and outlet channels, as well as channels for the two electrodes.

Experimental Procedure and Image Capturing

The two main variables tested for their impact on erythrocyte rupturing were age of the blood and blood type. It was important to hold the field density constant at .03 Vpp/micron for all experiments. This was done by measuring the gap distance between the electrodes and calculating the peak-to-peak voltage by:

$$E = {}^{V}\!/_{d}$$

where E is the constant field density of .03 volts/micron. The peak-to-peak voltage V was adjusted given the measured distance between the electrodes, d. Age experiments were determined from day 0 (the day the blood was drawn) and every two days thereafter.

A Zeiss Axiovert 200M inverted light microscope with a high resolution Axiocam camera was used to record images of the erythrocytes every fifteen seconds for fourteen minutes. The total number of remaining erythrocytes in the images were manually tabulated. All data was graphed using a percentage as a function of time:

$$X_{cells} = 100 * \frac{\#RBC_{time=t}}{\#RBC_{time=0}}$$

where the number of red blood cells at time zero was the initial amount of red blood cells visible in the field. These experiments were replicated twice a day every two days following donation. This allowed for a calculation of standard deviation for each blood type on Day 0.

Results

The number of ruptured cells were compiled and tabulated according to the formula above for the blood types A+, B+, AB+, O+ and O-. This was completed for Day 0, the day of donation and results compiled for the 14 minute run. A complete rupture profile was established for each blood type, as well as an overall percentage of red blood cells ruptured during the course of the experiment.

The percentage of red blood cells ruptured as a function of time for blood group A for one of the two Day 0 experiments is shown in Figure 2. For the first minute after the electric field is applied, the number of cells in the field of view increases due to cell migration into the imaging field of view. The cells then rapidly begin to rupture such that >30% rupture is achieved in the second minute. It is also interesting to note that after 300 seconds there seems to be no additional rupturing.

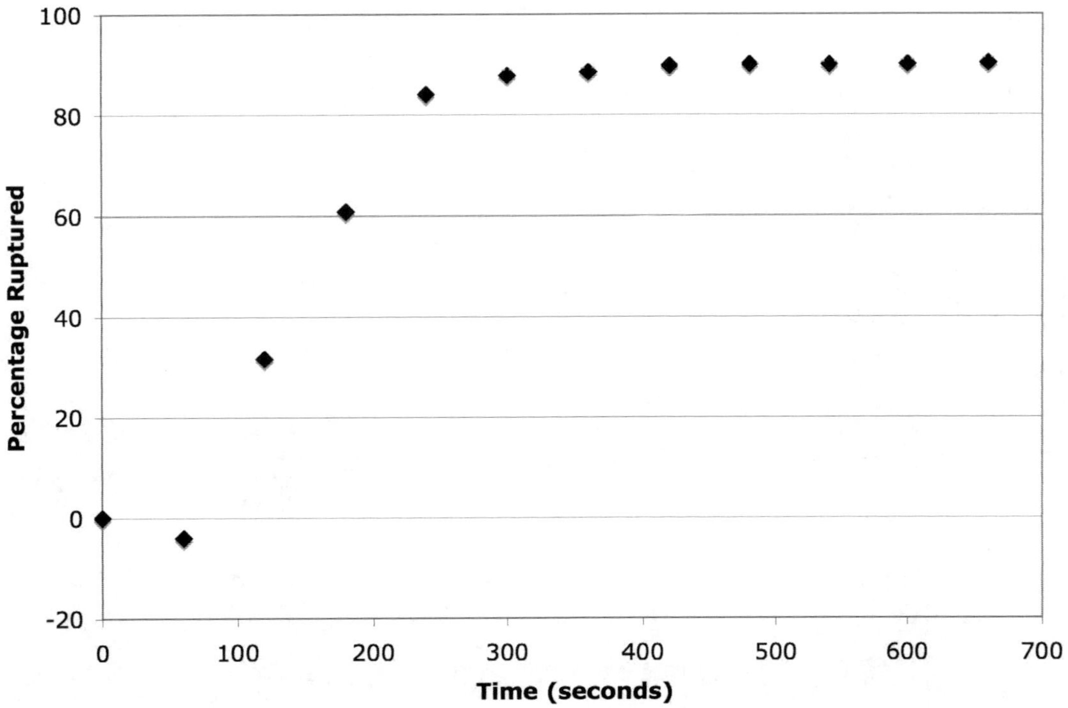

Figure 2: Rupturing of A+ on Day 0. The dip below 0% indicates a slight increase in the number of cells in the viewing area. Approximately 90% rupture is achieved in 660 seconds.

When comparing these profiles for each blood type, it became apparent that the rate of rupture varied greatly based on blood type for all of the Day 0 experiments. The percent of rupturing is compiled as a function of the blood type in Figure 3. For example, B+ red blood cells demonstrate minimal rupture over the course of 14 minutes, whereas approximately 50% of the O- red blood cells are ruptured. Standard deviation is not reported with A+ blood because while one experiment yielded bubbles at the electrodes (usually indicative of transient electrolysis reactions at the Platinum surface) and >90% rupture, the second experiment demonstrated no bubble formation and 0% rupture. The electrical connection was likely not complete in this second experiment and is in the process of being repeated. Also there are noted variations in the B+, O+ and O- results, which is hypothesized to be a result of the device design.

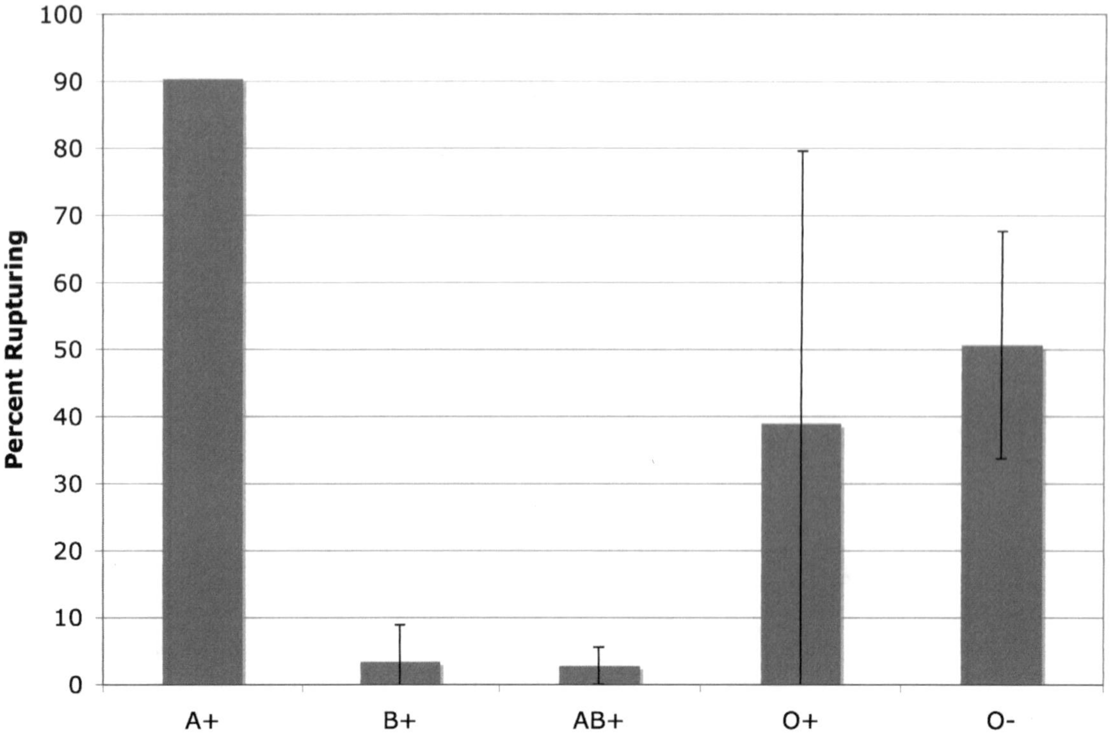

Figure 3: Percentage Rupturing over the Course of the Experiment. A+ ruptures more of its cells over the course of the experiment than any other blood type. Blood types containing B antigen (B+ and AB+) don't appear to rupture at all.

It is interesting to note that the presence or absence of the A and B antigens appear to have more of an effect on rupturing percentage than does the presence of the Rh antigen. This can be seen when comparing the large variation in rupturing percentages between A+, B+, AB+ and O+ but the comparatively similar rupturing percentages between O+ and O-.

Conclusions

Preliminary data from the influence of dielectrophoretic fields on various blood types suggests interesting trends. It was hypothesized that the blood type would have an effect on the rupturing rate due to their influence of antigens on membrane stability. Preliminary results suggest rupturing rate is influenced by blood type. It can be seen that the B+ blood barely ruptures whereas the O- blood ruptures approximately 50% of its cells in 14 minutes. The placement of the electrodes and their shape must be strictly controlled to keep an accurate field density of .03Vpp/micron for all of the experiments. The next step in this project is to quantify the affect of age on the rupturing rate of the red blood cells. This will involve completing analysis of experimental runs completed 2, 4 and 6 days after donation.

These experiments will further understanding the membrane properties of the red blood cells as they age. Understanding membrane properties of red blood cells will allow for the eventual implementation of dielectrophoretic rupturing onto lab-on-a-chip devices. The need to understand how these membrane properties change over time becomes very important when tests done in the field must be repeated later or when blood banks want to test blood in storage for its viability. These rupturing experiments give us important information about how various blood types respond to electric fields while aging.

References

[1] Hughes, M. "Strategies for dielectrophoretic separation in laboratory-on-a-chip systems." *Electrophoresis.* Vol. 23, pp. 2569-2582, 2002.

[2] Pohl, H. *Dielectrophoresis.* New York, NY: Cambridge University Press, 1978.

[3] Minerick, A.R., R. Zhou, P. Takhistov, and H.-C. Chang. "Manipulation and Characterization of Red Blood Cells with Alternating Current Fields in Micro-Devices" *Electrophoresis*, 24(21): 3703-3717, 2003.

[4] Pethig, R. "Dielectrophoresis: Using Inhomogeneous AC Electric Fields to Separate and Manipulate Cells." *Crit. Rev. Biotech.* Vol. 16(4), pp. 341-348, 1996.

[5] Srivastava, S. K.; Daggolu, P. R.; Burgess, S. C.; Minerick, A. R.; "Dielectrophoretic characterization of erythrocytes: Positive ABO blood types", *Electrophoresis* (in press, July 2008).

[6]Vissers, M.C.M. et al. *Biochem J.* Vol 330, pp. 131-138, 1998.

[7] Minerick, A.R., "The Rapidly Growing Field of Micro and Nanotechnology to Measure Living Cells," Invited Perspective with Cover Figure, *AIChE Journal*, Volume 54, Issue 9, Pages 2230-2237, September 2008.

[8] Menachery, A., and R. Pethig. "Controlling cell destruction using dielectrophoretic forces." *IEEE Proceedings.* 2005.

[9] Wang, J., Stine, M.J. and C. Lu. "Microfluidic Cell Electroporation Using a Mechanical Valve." *Anal. Chem.* Vol. 79, 9584-9587, 2007.

[10] Dailey, J.F. *Blood.* Arlington, MA: Medical Consulting Group, 1998.

[11] Landsteiner, K.Zur Kenntnis der antifermentativen, lytischen und agglutinierenden Wirkungen des Blutserums und der Lymphe. Zbl. Bakt. 27: 357-362, 1900.

[12] "Rhesus blood group system", Wikipedia. http://en.wikipedia.org/wiki/Rhesus_factor

[13] Leonard, Kaela, Reeves, Sheena, Pate, Ashley, Walton, Megan, Thompson, Sarah and Adrienne Minerick (November 2007) Rapid Cell Rupture in a Dielectrophoretic Field. Poster presented at 2007 Annual AES Conference.

[14] McDonald. J.C. et al. "Fabrication of microfluidic systems in poly(dimethylsiloxane)." *Electrophoresis.* Vol. 21, pp. 27-40, 1999.

65c Dielectrophoretically Induced Electrothermal Cell Separation

Zachary R. Gagnon[1], Satyajyoti Senapati[2], and H.-C. Chang[1]. (1) Chemical and Biomolecular Engineering, University of Notre Dame, 182 Fitzpatrick Hall, Notre Dame, IN 46556, (2) Chemical & Biomolecular Engineering, University of Notre Dame, 182 Fitzpatrick Hall, Notre Dame, IN 46556

Dielectrophoresis (DEP) is a term commonly used to describe the field induced polarization and translational motion of a polarizable particle in a non-uniform AC field. Traditional DEP detection, particle sorting and concentration schemes are typically based on the determination and exploitation of particle specific DEP crossover frequencies (cof).

It has been shown previously [1,2] that bovine red blood cells (bRBCs) exhibit two cof's, a low frequency (~200 kHz) cof being sensitive to the properties of the cell membrane, and the high frequency cof (~ 2 MHz) being sensitive to properties of the cell cytoplasm. This dual cof behavior is due to the shell-like nature of the RBC and hence allows one to analyze and separate cells based on their unique membrane and interior properties.

In this work we carry the above analysis further using human red blood cells (hRBCs). We show that due to differences in membrane and cytoplasm properties hRBCS have largely different DEP behavior than bRBCs. By utilizing a glutaraldehyde crosslinking cell fixation reaction that is sensitive to cell membrane protein concentration, we demonstrate how species-specific protein and lipid cell membrane surface concentrations affect the dielectric response of blood. Our results reveal that, after cross-linking to produce conductive and polarized double bonds along their backbones, the lipid and amino-terminal protein concentrations in the cell membrane are essentially what determine the membrane conductivity and dielectric constant.

Such observations are exploited in order to separate a suspension of malaria infected red blood cells (~ 5% infected) from their healthy counterpart. A novel closed-loop electrothermal micro-pump is fabricated such that approximately 20 nL of infected blood is forced to circulate around a closed loop. By controlling the separation distance of the working electrodes, the device is engineered such that the field penetration (20 ìm) does not exceed the height of the microchannel (75 ìm). By crosslinking the infected cell suspension it becomes possible to separate healthy cells based on the reaction-induced differences in DEP mobility. It is shown by selectively staining the infected cells that healthy cells are concentrated at specific high field trapping regions, while infected cells are forced into fluid streamlines above the trap and swept away to continue around the closed loop via electrothermal fluid flow.

[1] Z. Gagnon, J Gordon, S Sengupta and H.-C. Chang Electrophoresis 2008, 29 (In press)

[2] Gordon, J.G., Gagnon, Z., Chang, H.-C., Biomicrofludics 2007, 1, 044102

Capillary Electrophoresis of PEGylated Proteins

Dr. Jessica R. Molek, GlaxoSmithKline, King of Prussia, PA
Dr. Andrew L. Zydney, The Pennsylvania State University, University Park, PA

Introduction

Capillary electrophoresis has become a well-established technique for evaluating the composition and concentrations of a wide range of protein samples[1-4]. In addition, capillary electrophoresis can be used to study a variety of protein properties including the net protein charge[5], the extent of ion binding[6], and the presence of specific biomolecular interactions[7-9]. This wide range of applications is possible, at least in part, because of the strong theoretical foundation for describing the electrophoretic mobility of globular proteins. Smoluchowski's initial analysis of the electrophoretic mobility is valid for arbitrary particle shape under conditions where the Debye length (double layer thickness) is much smaller than the characteristic particle size. Subsequent studies have accounted for the finite size[10] and relaxation of the electrical double layer[11], the effects of non-spherical particle geometry[12], and the influence of particle porosity[13-16]. The resulting theoretical expressions for the electrophoretic mobility are in good agreement with experimental results. In contrast, much less is known about the electrophoretic mobility of more complex biomolecules. For example, there is considerable interest in the use of PEGylated therapeutic proteins, formed by the covalent attachment of one or more polyethylene glycol chains to the native protein. PEGylation can significantly increase the biological half-life by reducing the rate of kidney clearance[17], and it can also minimize adverse immunological reactions[18], dramatically improving the overall efficacy of the therapeutic protein.

Fundamental studies of PEGylated proteins have clearly demonstrated that the attachment of the PEG chain(s) increases the effective protein size as determined by size exclusion chromatography[19, 20] and membrane sieving[20]. PEGylation also increases protein hydrophobicity, as determined by reverse phase chromatography[21], and it significantly reduces dynamic binding capacity in ion exchange chromatography [22], although no detailed models have been presented for either of these effects. PEGylated proteins do retain their electrical charge, with the resulting electrostatic interactions leading to a significant increase in retention during ultrafiltration through electrically-charged membranes with like polarity[23]. In addition, there is evidence that the polyethylene glycol chains on PEGylated proteins can elongate under flow, reducing the effective protein size at high flow rates[20].

Several investigators have used capillary electrophoresis for analysis of PEGylated protein samples. Semi-aqueous capillary electrophoresis has been used to resolve the different PEGylated forms of ribonuclease A[24], lysozyme[25], and recombinant TNF-α. Li et al.[24] used capillary electrophoresis to measure the extent of aggregation of PEGylated ribonuclease A under different storage conditions. More recently, Na et al.[27] used SDS capillary gel electrophoresis to check the purity of PEGylated α-interferon with both single and multiple PEG chains. In addition, Sharma and Carbeck[28] used capillary electrophoresis to calculate the effective size of PEGylated lysozyme, implicitly assuming that the electrical charge on the native protein is displaced to the outer radius of a PEGylated sphere. The calculated values of this effective radius were in good agreement with results from partitioning in size exclusion chromatography, although the authors provided no physical basis for the use of this model.

The objective of this study was to obtain quantitative data for the electrophoretic mobility and resolution of PEGylated α-lactalbumin using capillary electrophoresis. Data were also obtained with an acetylated form of α-lactalbumin, which was generated by covalent modification of the lysine amino groups but without any significant change in the protein molecular weight due to the small size of the ligand. Data were analyzed using available theoretical models for the electrophoretic mobility of composite objects to obtain additional insights into the underlying structure of the PEGylated proteins.

Materials and Methods

All experiments were performed using α-lactalbumin as a model protein (Sigma Chemical, St. Louis, MO). The protein was PEGylated with mPEG-SPA (activated with succinimidyl propionic acid) or mPEG-SMB (activated with succinimidyl α-methylbutanoate), each having nominal weight-average molecular weight of 2, 5, 10, 20, or 30 kDa. PEGylation was performed by slowly adding the activated PEG to a 10 g/L solution of α-lactalbumin in 10 mM Bis-Tris buffer at pH 7. Acetylated proteins were synthesized using the general procedure described by Gao and Whitesides[5], which converts the free amine groups on one or more lysines to the corresponding (uncharged) amide.

The electrophoretic mobility of the PEGylated and acetylated proteins were determined using an Agilent G1600A high performance capillary electrophoresis system equipped with a dual-polarity variable high voltage DC supply (0-30 kV) and a diode array UV/visible absorbance detector (214 nm wavelength for detection). Experiments were performed with negatively-charged fused-silica capillaries (inner diameter of 0.5 μm) with a total length of 80.5 cm and an effective length (from injection to the detection window) of 70.2 cm. The capillary and solution reservoirs were filled with buffered electrolyte, most commonly either 10 mM Tris/Glycine (pH 8.1) or 10 mM Bis Tris (pH 7.0). Protein samples (approximately 40 - 140 nL) containing 5 mM of mesityl oxide as a neutral marker were injected for 4 to 25 s at a pressure of 4 kPa. Data were obtained at a constant applied voltage of 25 kV, with the field direction chosen so that the bulk electroosmotic flow was toward the detector (and cathode). The electrophoretic mobility was calculated from the migration times as:

$$\mu_e = \frac{L_d}{E_z}\left(\frac{1}{t} - \frac{1}{t_o}\right) \qquad (1)$$

where L_d is the effective capillary length of capillary to the detector, E_z is the applied electric field, and t and t_o are the times for the protein and neutral marker to reach the detector.

Results and Discussion

Several different running buffers were examined to obtain good resolution and sharp symmetric peaks. Previous studies have shown that the buffer type and composition can strongly influence the overall resolution and separation[29]. Initial experiments used the semi-aqueous (water + acetonitrile) buffer employed by Li et al.[30] for analysis of PEGylated RNA-ase. Results were obtained between 10 and 80% acetonitrile, but peak resolution was poor and decreased at high acetonitrile concentrations. Typical results using 10 mM Bis Tris buffer (the buffer used in the PEGylation reaction) are shown in Figure 1 (lower electropherogram). The first peak is the neutral marker (mesityl oxide) that was added to the PEGylated sample. The second distinct peak is the PEGylated α-lactalbumin with one 20 kDa branch. The third peak is the native α-lactalbumin. Although it was possible to resolve peaks corresponding to the PEGylated proteins with different numbers of PEG chains and the native protein, there was significant peak asymmetry and the overall resolution was poor using the Bis Tris buffer. The upper electropherogram in Figure 1 shows typical results using a 10 mM Bis Tris with 0.1 M glycine added to the buffer. The resolution between α-lactalbumin and the PEGylated α-lactalbumin with a single 20 kDa PEG chain increased from 2.1 to 4.7 upon addition of the glycine, consistent with results obtained previously using other zwitterionic additives[2]. In addition, the symmetry factor of the α-lactalbumin peak decreased from 13.6 to 3.6 in the presence of glycine. The improved resolution is likely due to competition of the glycine for the silanol groups on the capillary, which reduces the extent of protein adsorption, along with the suppression of joule heating associated with zwitterions[31].

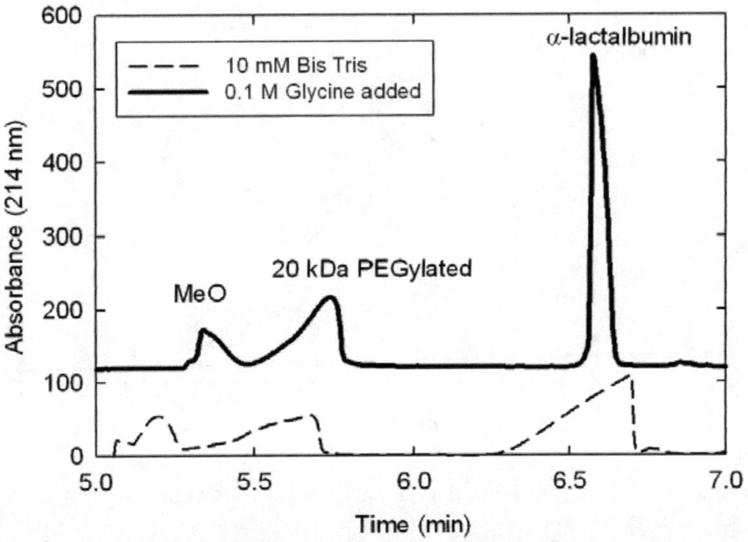

Figure 1. Electropherogram of a 20 kDa PEGylated protein with running buffers of Bis Tris (pH 7.0, 10 mM) (lower curve), and Bis Tris with 0.1 M glycine (pH 7.0, 11mM) (upper curve).

Figure 2 shows electropherograms for both the acetylated (lower curve) and PEGylated (upper curve) α-lactalbumin using a 10 mM pH 8.1 Tris-Glycine running buffer at 25 kV. In both cases, the first peak is the neutral marker (mesityl oxide), which is carried to the detector solely by the electroosmotic flow generated within the negatively charged capillary. The second peak with the acetylated sample corresponds to the native α-lactalbumin, which migrates back against the electroosmotic flow and thus exits at a later time than the neutral marker. Each subsequent peak in the lower curve represents an acetylated α-lactalbumin with one more lysine group capped by reaction with acetic anhydride, resulting in a more negatively charged species. Twelve distinct peaks were easily resolved, with only the first four shown in Figure 2. Note that even the most heavily acetylated species in this system has a molecular weight that is only 3.6% larger than the unmodified α-lactalbumin. The upper curve in Figure 2 shows the electropherogram for the PEGylated α-lactalbumin formed by reaction with 5 kDa mPEG-SPA. In contrast to the data for the acetylated α-lactalbumin, the peaks for the PEGylated species lie between those for the neutral marker and the unmodified α-lactalbumin due to the reduction in the electrophoretic mobility associated with the addition of the large neutral polyethylene glycol chain(s). Three distinct peaks, each corresponding to a PEGylated species with one more PEG chain, are easily observable in the electropherogram. A fourth peak, corresponding to α-lactalbumin PEGylated with four PEG chains, appears as a shoulder at $t = 6.22$ min. The peaks for the PEGylated species are much broader than those for the acetylated α-lactalbumin, due primarily to the polydispersity in the mPEG-SPA.

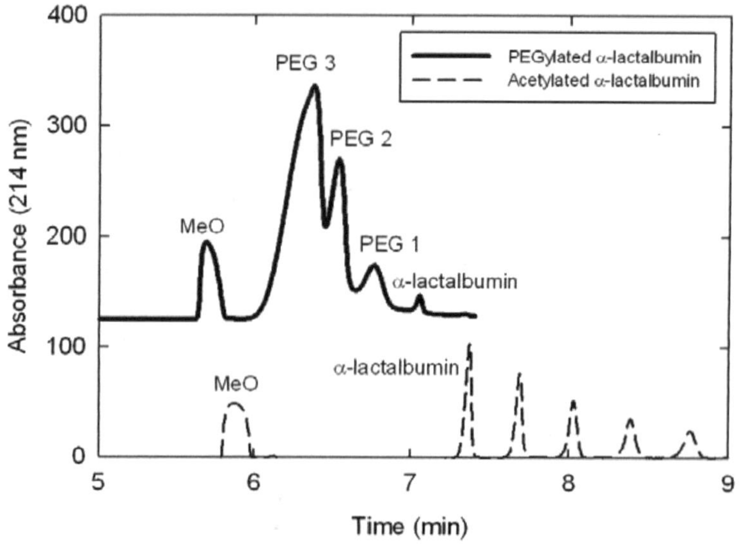

Figure 2. Electropherograms of acetylated α-lactalbumin (lower curve) and α-lactalbumin PEGylated with 5 kDa mPEG-SPA (upper curve).

The data in Figure 2 were used to calculate the electrophoretic mobility of the acetylated and PEGylated species using Equation 1. Results are shown in Figure 3 for experiments performed with PEGylated α-lactalbumin formed by attachment of PEG with molecular weight of 2, 10, 20, and 30 kDa PEG along with the corresponding results for the acetylated α-lactalbumin. In each case, the data are plotted as a function of the number of modifications to the native protein; thus, the species with n = 1 represents α-lactalbumin with either a single PEG group or a single acetylated lysine. The electrophoretic mobility of the acetylated species increases with increasing number of acetyl groups due to the increase in net negative charge associated with the conversion of the protonated amino group to an amide. In contrast, the electrophoretic mobility of the PEGylated α-lactalbumin decreases with increasing number of attached PEG groups even though PEGylation eliminates multiple amine groups (analogous to the acetylation reaction). The electrophoretic mobility of the PEGylated α-lactalbumin with one 20 kDa PEG chain is slightly smaller than that of the PEGylated protein with two 10 kDa PEG chains which is turn smaller than that for the PEGylated protein with four 5 kDa PEG chains even though these species all have basically the same molecular weight (and effective size as determined by size exclusion chromatography[19]). These differences are a direct result of the change in molecular charge, in addition to size, due to PEGylation.

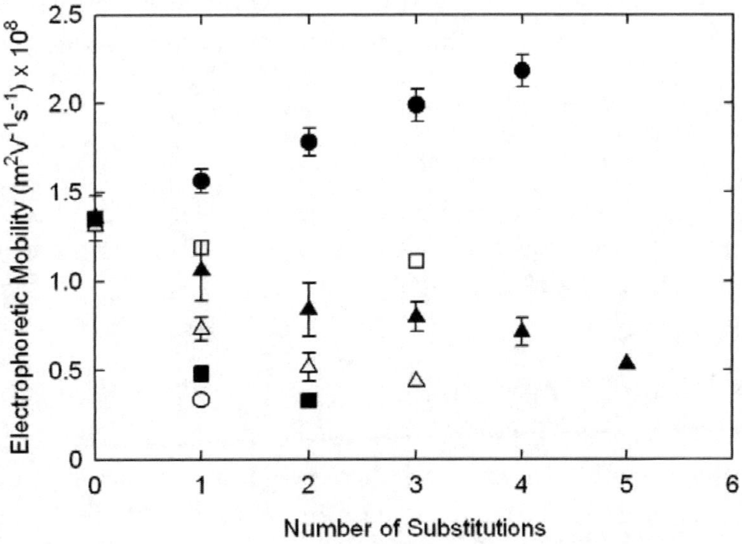

Figure 3. Electrophoretic mobility of the acetylated α-lactalbumin (•) and PEGylated α-lactalbumin with 2 kDa branches (□), 5 kDa branches (▲), 10 kDa branches (Δ), 20 kDa branches (■) and 30 kDa branches (○) as a function of the number of substituted lysine groups.

In order to account for the change in both size and net charge of the PEGylated proteins, the experimental data in Figure 3 have been analyzed using the drag ratio, K_d, which is equal to the electrophoretic mobility of the PEGylated α-lactalbumin divided by the mobility of the corresponding acetylated species with the same number of chemical substitutions. Thus, the drag ratio for the PEGylated α-lactalbumin with a single 10 kDa PEG group was evaluated by normalizing the mobility using data for the singly-acetylated protein while the drag ratio for the PEGylated α-lactalbumin with two 5 kDa PEG groups was evaluated by normalizing with data for the doubly-acetylated protein. This method effectively eliminates the contribution of charge since the PEGylated and acetylated proteins have the same number and distribution of charged amino acid residues. Sharma and Carbeck[28] used a similar approach to calculate the effective size of PEGylated proteins using capillary electrophoresis.

Experimental data for the drag ratio for the different PEGylated species are shown in Figure 4 as a function of the effective protein radius, which was calculated using the correlation presented by Fee and Van Alstine[32] based on partitioning data in size exclusion chromatography:

$$a_{PEGylatd} = \frac{C}{6} + \frac{2}{3C} a_{PEG}^2 + \frac{1}{3} a_{PEG} \qquad (2)$$

$$C = \left[108a_{pro}^3 + 8a_{PEG}^3 + 12\left(81a_{pro}^6 + 12a_{pro}^3 a_{PEG}^3\right)^{1/2} \right]^{1/3} \qquad (3)$$

where a_{pro} and a_{PEG} are the radii of the unmodified protein and the free PEG chain, respectively. The radius of a free PEG molecule was calculated as:

$$a_{PEG} = 0.01912 \times MW^{0.559} \qquad (4)$$

where a_{PEG} is in nm and the molecular weight (MW) is in Da[33]. The data in Figure 4 collapse to a single curve when plotted in this manner. For example, the drag ratio for the PEGylated α-lactalbumin with a single 10 kDa PEG chain is 0.44, which is within 10 % of the value for the PEGylated α-lactalbumin having two 5 kDa PEG chains (which is within the standard deviation of the measurements). This is in sharp contrast to the 40% difference in the electrophoretic mobility of these species, which is directly due to the difference in electrical charge associated with the different number of modified lysine groups.

The solid curve in Figure 4 represents the predicted drag ratio calculated using Henry's equation:

$$\frac{\mu_e}{\mu_{e,o}} = \left[\frac{f\left(\kappa a_{PEGylated}\right)}{f\left(\kappa a_{\Pr o}\right)}\right]\left(\frac{a_{\Pr o}}{a_{PEGylated}}\right)\left(\frac{1+\kappa a_{\Pr o}}{1+\kappa a_{PEGylated}}\right) \tag{5}$$

Equation (5) assumes that the PEGylated protein can be treated as a hard-sphere with net electrical charge equal to that of the acetylated species having the same number of modifications. The radius of the unmodified α-lactalbumin is $a_{pro} = 1.98$ nm while the radius of the PEGylated protein was evaluated using Equations (2) – (4). Henry's function, $f(\kappa a)$, was evaluated using the following expansion in κa[10]:

$$f(\kappa a) = 1 + \frac{1}{16}(\kappa a)^2 - \frac{5}{48}(\kappa a)^3 - \frac{1}{96}(\kappa a)^4 + \frac{1}{96}(\kappa a)^5 + \frac{1}{8}(\kappa a)^4 \exp[\kappa a]\left(1 - \frac{(\kappa a)^2}{12}\right)\int_{\kappa a}^{\infty}\frac{\exp[-t]}{t}dt \tag{6}$$

where κ is the inverse Debye length. The model is in very good agreement with the experimental data, properly capturing the reduction in electrophoretic mobility caused by the increase in drag arising from the polyethylene glycol chains, with the change in protein charge associated with the PEGylation reaction incorporated through the normalization of the data using the electrophoretic mobility of the acetylated α-lactalbumin having the same number of substitutions.

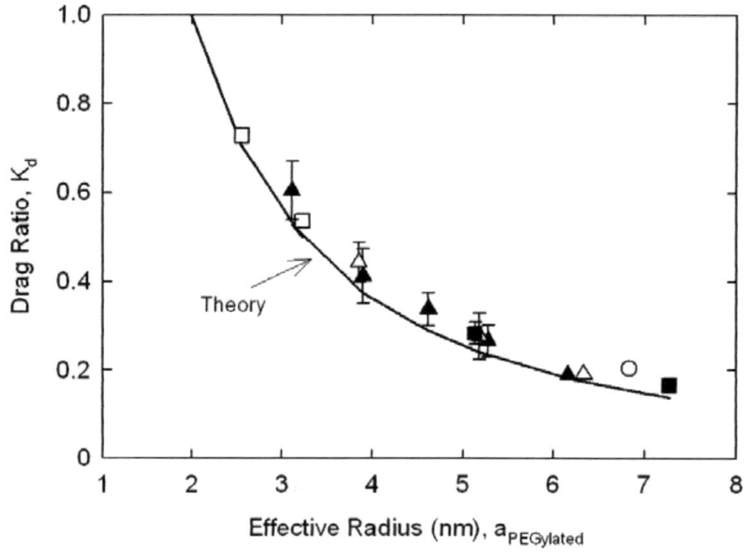

Figure 4. Drag ratio for PEGylated proteins as a function of the effective protein radius as shown with branches of 2 kDa PEG (□), 5 kDa branches (▲), 10 kDa branches (Δ), 20 kDa branches (■) and 30 kDa branches (○).

The good agreement between the experimental data and this simple model is in some ways surprising. Henry's analysis implicitly assumes that the electrical charge is distributed uniformly over the outer surface of a hard sphere, while the charge groups on the PEGylated proteins are distributed over the surface of the protein, which should be at least partially "buried" beneath a layer of polyethylene glycol. One possible physical explanation for this behavior is that the polyethylene glycol forms a shell that has a substantially lower ionic strength than in the bulk solution. The thermodynamics of PEG-salt systems have been studied quite extensively[34-37]. These systems tend to phase separate due to the strong "negative" interactions between the salts and the PEG. The salt concentration in the PEG phase can be as much as seven times smaller than the salt concentration in the non-PEG phase[34]. Although it is difficult to directly extrapolate these data for PEG-salt systems to the behavior of PEGylated proteins, the results clearly indicate that there may be a significant exclusion of buffer salts from the polyethylene glycol layer in the PEGylated protein. Alternatively, the polyethylene glycol chain(s) may form a "tail" behind the moving protein, with the electrical force governed by the charge on the actual surface of the protein while the drag is determined by the sum of the drag forces on the protein and polyethylene glycol "tail". Theoretical calculations using a model developed by Long and Ajdari[38] based on this physical situation gave values of the drag ratio that were much smaller than those determined experimentally, although some of the key assumptions in this model are unlikely to be valid in our experimental system .

Conclusions

The addition of glycine significantly improved the peak symmetry and resolution for the capillary electrophoresis of PEGylated proteins with different numbers of attached PEG chains. PEGylation alters the electrophoretic mobility by increasing the hydrodynamic drag (due to the presence of the PEG chains) and by shifting the net protein charge (through reaction with the $-NH_2$ of the lysine). The experimental data for the electrophoretic mobility of the different PEGylated species were collapsed to a single curve by plotting the drag ratio, equal to the electrophoretic mobility of the PEGylated α-lactalbumin divided by the mobility of the acetylated protein having the same number of covalently linked lysines, as a function of the effective radius of the PEGylated protein. In addition, the drag ratio data were in very good agreement with model predictions based on Henry's equation assuming that the net charge on the surface of the PEGylated protein is the same as that on the surface of the acetylated protein, despite the presence of the polyethylene glycol layer. This unusual behavior may reflect the exclusion of charged ions from the polyethylene glycol layer, a phenomenon that could have important implications for understanding the transport and separation characteristics of PEGylated proteins.

References

1. Carbeck, J. D. et al. *Phys Chem B* **1998**, *102*, 10596-10601.
2. Corradini, D. *J Chromatogr B* **1997**, *699*, 221-256.
3. Na, D. H. et al. *J Chromatogr B* **2001**, *754*, 259-263.
4. Watzig, H. et al. *Electrophoresis* **1998**, *19*, 2695-2752.
5. Gao, J.; Whitesides, G. M. *Anal Chem* **1997**, *69*, 575-580.
6. Menon, M.; Zydney, A. L. *Anal Chem* **1998**, *70*, 1581-1584.
7. Martinez-Gomez, M. A. et al. *J Chromatogr A* **2007**, *1147*, 261-269.
8. Caravella, J. A. et al. *J Am Chem Soc* **1999**, *121*, 4340-4347.
9. Safi, S.; Asfari, Z.; Hagege, A. *J Chromatogr A* **2007**, *1173*, 159-164.
10. Henry, D. C. *Proc R Soc Long Ser - A* **1931**, *133*, 106-129.
11. O'brien, R. W.; White, L. R. *J Chem Soc - Faraday Trans II* **1978,** 74, 1607-1626.
12. Hsu, J.-P.; Hung, S.-H. *J Colloid Interf Sci* **2003**, *264*, 121-127.
13. Ding, J. M.; Keh, H. J. *Langmuir* **2003**, *19*, 7226-7239.
14. Hill, R. J. et al. *J Colloid Interf Sci* **2003**, *258*, 56-74.
15. Cohen, J. A.; Khorosheva, V. A. *Colloid Surface A* **2001**, *195*, 113-127.
16. Bhatt, B. S.; Sacheti, N. *J Phys D Appl Phys* **1994**, *27*, 37-41.
17. Chapman, A. *Adv Drug Deliver Rev* **2002**, *54*, 531-545.
18. Edwards, C. K. et al. *Adv Drug Deliver Rev* **2003**, *55*, 1315-1336.

19. Fee, C. J.; Van Alstine, J. M. *Bioconjugate Chem* **2004**, *15*, 1304-1313.
20. Molek, J.; Zydney, A. L. *Biotechnol BioEng* **2006**, *95*, 474-482.
21. Kwon, B.; Molek, J. R.; Zydney, A. L. *J Membrane Sci* **2008**, *319*, 206-213.
22. Pabst, T., M. et al. *J Chromatogr A* **2007**, *1147*, 172-182.
23. Molek, J. R.; Zydney, A. L. *Biotechnol Progr* **2007**, *23*, 1417-1424.
24. Li, W.; Su, Z. *J Biochem Bioph Meth* **2004**, *59*, 65-74.
25. Corradini, D et al. *Chromatographia Supplement* **2005**, *62*, S43- S50.
26. Li, W. et al. *J Biotechnol* **2002**, *92*, 251-258.
27. Na, D. H. et al. *Anal Biochem* **2008**, *373*, 207-212.
28. Sharma, U.; Carbeck, J. D. *Electrophoresis* **2005**, *25*, 2086-2091.
29. Shihabi, Z. *J Chromatogr A* **2004**, *1027*, 179-184.
30. Li, W. et al. *J Chromatogr A* **2001**, *905*, 299-307.
31. Bushey, M. M.; Jorgenson, J. W. *J Chromatogr A* **1989**, *480*, 301-310.
32. Fee, C. J.; Van Alstine, J., M. *Chem Eng Sci* **2006**, *61*, 924-939.
33. Singh, S. et al. *J Membrane Sci* **1998**, *142*, 111-127.
34. Willauer, H. D. et al. *Ind Eng Chem* **2002**, *41*, 1892-1904.
35. de Belval, S. et al. *J Chromatogr B* **1998**, *711*, 19-29.
36. Andrews, B. A. et al. *Biotechnol BioEng* **2005**, *90*, 380-390.
37. Schmidt, A. S. et al. *Biotechnol BioEng* **1996**, *50*, 617-626.
38. Long, D.; Ajdari, A. *Electrophoresis* **1996**, *17*, 1161-1166.

65e Effect of Serum Contaminants on the Detection of DNA and RNA Using Surfactant Probes in Capillary Electrophoresis

Oxana Selivanova and James W. Schneider. Chemical Engineering, Carnegie Mellon University, 700 Technology Drive, Room 4307, Pittsburgh, PA 15219

Quantifying DNA and RNA sequences is a key process in the analysis of gene expression and identification of microorganisms. Gene expression quantification such as microarray methods is necessary for tracking the levels of messenger RNA (mRNA) in cells. mRNA purification from cells requires multiple steps due to different cellular constituents such as proteins and lipids present. Significant savings in cost and time could be realized if levels of DNA and RNA are measured in minimally processed samples, which would contain significant amounts of proteins, lipids, and other contaminants.

Micellar Electrokinetic Chromatography (MEKC) employs a surfactant probe that transiently interacts with micelles via its hydrophobic tail. Uncharged micelles migrate with electroosmotic flow and speed up the mobility of the probe. The resultant mobility of the probe is the average between its free solution mobility and micelle mobility. The probe sequence specifically hybridizes to complementary DNA/RNA shifting its mobility thus allowing quantification of the desired nucleic acid. Cellular contamination such as serum proteins can adsorb to the walls of bare silica capillaries thus requiring employment of coated capillaries. Proteins also result in micelle swelling. We are interested in the maximum amount of serum proteins that can be used with bare silica capillary for the successful DNA/RNA quantification. Alkylated DNA (aDNA) surfactant probe sequence specifically hybridizes to single stranded (ss) DNA creating aDNA/DNA duplex which partitions into MEKC micelles. The resultant output is an electropherogram where DNA is quantified via duplex mobility shift and peak intensity. Proteins are known to interact with DNA via G-C bases and DNA phosphodiester backbone. These interactions lead to the emergence of multiple and distorted peaks. Our results suggest that the probe prevents DNA from protein interaction. The position of overhanging DNA bases is crucial because hindered by the hydrophobic tail bases show no protein-DNA interaction. Only two peaks appear corresponding to aDNA and aDNA/DNA duplex as judged by a control without goat serum. Goat serum under 1mg/mL has no effect on DNA detection. At 2mg/mL significant peak distortion possibly due to protein adsorption onto bare silica capillary occurs. The next step is to detect longer ssDNA which is representative of cellular mRNA.

65f Capillary Channels with Electroosmotic Flows: Role of Fundamental Forces In Promoting Flow Reversal

Karina Merino[1], **Mario Oyanader**[2], and Pedro E. Arce[2]. (1) Chemical Engineering, University Catolica del Norte, Avda Angamos 0610, Antofagasta, Chile, (2) Chemical Engineering, Tennessee Technological University, Department of Chemical Engineering, P.O. Box 5013, Cookeville, TN 38505

Soil clean up applications by electrokinetic process have demonstrated to be suitable for heavy metal removal in sites of difficult access as an in situ technique. The effect of different driving parameters has been identified and reported in the literature as an effort to unveil the optimal conditions of removal. Among others, pH profile variations along the treatment zone have been linked as an important parameter to process efficiency. Studies have indicated that under certain condition axial flow reversals occurs affecting contaminant removal. Validating this assumption is important as well as identifying the implicated results. In this study a particulate movement is simulated for different case scenarios. The primary mechanisms considered in the approach are electroosmosis and drag forces. The theoretical analysis has been closely monitored and compared with laboratory results. Meaningful illustrative results are presented along with the most relevant conclusions. The idea behind the study is to explain the experimental results via theoretical modeling and simulation.

65g Electrofocusing of Trace Contaminants

Jeffrey M. Burke and Cornelius F. Ivory. School of Chemical and Biological Engineering, Washington State University, Pullman, WA 99163

The analysis of impurities in pharmaceutically active compounds is becoming increasingly important as federal regulations become stricter. The most common technique for liquid phase analysis of contaminants is high performance liquid chromatography (HPLC). In general, however, only moderate resolution separations can be achieved and in many instances a prior concentrating technique is required for dilute samples. Isoelectric focusing (IEF), another commonly used liquid phase technique, separates and concentrates in the presence of a pH gradient and thus would not require a prior concentrating step. However, this technique requires that the analytes be amphoteric with an isoelectric point that is neither highly acidic nor highly basics. This is not always the case for pharmaceutics and thus could not be used for all samples. This paper discusses the use of dynamic field gradient focusing (DFGF) for the removal of trace contaminants from a desired product. DFGF is an additional electrofocusing method that can simultaneously separate and concentrate charged analytes in the presence of an electric field gradient. Unlike isoelectric focusing which separates in the presence of a pH gradient, DFGF does not require that the analytes be amphoteric, only that they have differences in their individual electrophoretic mobilites. In DFGF the electric field gradient is generated by a computer-controlled electrode array that allows for dynamic control and manipulation of the field profile during the course of an experiment to increase peak resolution, migrate analytes to off-take ports or to systematically elute individual species. Experimental results from the separation of two low molecular weight dyes, bromophenol blue and amaranth, at a concentration difference of 1000 fold are shown. These results along with a numerical simulation constructed in Comsol Multiphysics v3.4 are used to illustrate the potential of DFGF for trace contaminant analysis.

136a DNA Electrophoresis in Sparse PDMS Micropillar Arrays

Jia Ou[1], Daniel Olson[1], Michael Meloche[2], and **Kevin D. Dorfman**[1]. (1) Department of Chemical Engineering and Materials Science, University of Minnesota, 151 Amundson Hall, 421 Washington Ave SE, Minneapolis, MN 55455, (2) Department of Biomedical Engineering, University of Minnesota, 151 Amundson Hall, 421 Washington Ave SE, Minneapolis, MN 55455

The separation of long DNA by electrophoresis through regular arrays of micropillars is a major accomplishment of microfluidics, with the potential of reducing day-long macroscale separations into the minute range. We will present new experimental data obtained for the transport properties of lambda-phage DNA (48,500 base pairs) in model post arrays. The devices are fabricated in PDMS, featuring centimeter-long hexagonal arrays of 1 micron posts with pitches of 3, 5 and 7 microns. We have implemented a robust method for fabricating and preparing these chips that allows for high chip-to-chip reproducibility and reusability. We will present results on the mobility and dispersivity of lambda-DNA as a function of the pitch of the array and the electric field. These data will be compared with prevailing theories of DNA electrophoresis, including continuous-time random walk predictions and the so-called "channeling" effect that is predicted by simulations of DNA electrophoresis in sparse, ordered arrays.

136b Influence of Pore Morphology on DNA Separation Performance In Microchip Gel Electrophoresis

Nan Shi, Roger C. Lo, Jian Wang, and Victor M. Ugaz. Artie McFerrin Department of Chemical Engineering, Texas A&M University, 3122 TAMU, College Station, TX 77843

Gel electrophoresis is an essential analytical step in a wide range of DNA analysis assays. Incredible progress continues to be made toward adapting electrophoresis technology for use in portable low-cost applications, but miniaturization demands increasingly precise control over all aspects of the process in order for separations to be performed in ultra-short distances. Here we explore the role of the sieving gel's nanoscale pore morphology in directing DNA migration during electrophoresis. We accomplish this by harnessing a unique combination of methods that enable both the mean pore size and the pore size distribution of the gel to be quantitatively characterized, and a versatile microfluidic platform that allows the progress of DNA separations to be continuously monitored along the entire microchannel in near real time. This approach allows us to identify how polymerization conditions (i.e., UV intensity during photopolymerization) influence the gel pore size distribution, and ultimately shape overall separation performance. Analysis of double-stranded DNA separations in the size range below 1 kb reveals that varying the rate of photopolymerization induces a corresponding change in the physical mechanism of DNA migration between reptation and entropic trapping. These results highlight the direct link between polymerization conditions, gel pore morphology, and DNA separation performance. We hope these measurements can help provide new insights into the interplay between the nanoporous architecture of the gel matrix and the electrophoretic transport process.

136c Using Optical Tweezers for Electroporation Process Analysis

Brian E. Henslee[1], Shengnian Wang[2], Xin Hu[3], Andrew Morss[4], L. James Lee[5], and Gregory Lafyatis[4]. (1) Department of Chemical and Biomolecular Engineering, The Ohio State University, Columbus, OH 43210, (2) Institute for Micromanufacturing, Louisiana Tech University, 911 Hergot Avenue, Ruston, LA 71272, (3) NSEC center for Affordable Nanoengineering of Polymer Biomedical Devices (CANPBD), The Ohio State University, Room 125A, 140 West 19th Ave., Columbus, OH 43210, (4) Department of Physics, The Ohio State University, Columbus, OH 43210, (5) Chemical and Biomolecular Engineering, The Ohio State University, 125 Koffolt Labs, 140 W 19th Ave, Columbus, OH 43210

Electroporation is widely used for drug and gene delivery with the aid of electric pulses. However, its mechanism is not fully discovered, and as a result, trial-and-error optimization process is required to identify proper electric conditions for every cell type system. By using optical tweezers technology, we are able to move and affixed single or multiple cells at desirable locations during electroporation. This provides the opportunity for in situ monitoring and quantifying the electroporation process at single or multiple cell level. We investigate the process for cell membrane polarization and pore formation as well as the uptake of therapeutic molecules and its intracellular transport. DNA dynamics and DNA-cell interactions are also quantified under different conditions, such as the relative positions between cells, cell to electrodes, between electrodes and the geometry and positions of electrodes. The possible influence of laser on the cell and cell membrane is isolated from electroporation and quantified. Our experimental observations are used as input for FEM models of cell electroporation and preliminary comparison is made. This study provides a new route to investigate the fundamental mechanism of nucleic acid transfer and intracellular response in electroporation. It is highly valuable to eliminate or simplify the optimization process of electroporation conditions for a new cell type, especially primary cells and stem cells.

136d Effect of the Electric Field on a Multi-Cell System in Complex Geometry

Xin Hu[1], Andrew Morss[2], Brian E. Henslee[3], Wei-Ching Liao[4], Shengnian Wang[5], Gregory Lafyatis[2], and Ly James Lee[3]. (1) NSEC center for Affordable Nanoengineering of Polymer Biomedical Devices (CANPBD), The Ohio State University, Room 125A, 140 West 19th Ave., Columbus, OH 43210, (2) Department of Physics, The Ohio State University, Columbus, OH 43210, (3) Department of Chemical and Biomolecular Engineering, The Ohio State University, Columbus, OH 43210, (4) Mechanical Engineering, The Ohio State University, Columbus, OH 43210, (5) Institute for Micromanufacturing, Louisiana Tech University, 911 Hergot Avenue, Ruston, LA 71272

As an effective non-viral method in gene delivery, cell electroporation applies an external electric field to temporarily increase the permability of the cell membrane. Currently, there are two types of cell electroporation, i.e., the bulk and single cell electroporations. The efficiency of the single cell electroporation is higher because it can precisely control the electric field distribution around the single cell. But the effect of electric field on a multi-cell system hasn't been fully understood yet, especially when the system is in a complex geometry. We use both experiment and simulation to study the effect of external electric field on a system with two or more cells in a complex contraction/expansion microchannel, which can focus the electric field near the area of smallest cross-section. By using charged tracing particles, we are able to draw the electric field lines in this multi-cell system. This also enables us to calculate the value of the electric conductivity of the cell. Both the experiment and simulation show that the distribution of the electric field varies when the relative position of cells changes. This study can be used to guide us to increase the efficiency of the bulk electroporation.

136e Characterization of Microfluidic Devices for Cell Separation Via Adhesion to Peptide-Functionalized Surfaces

Melissa A. Brown[1], Shashi K. Murthy[2], and Milica Radisic[1]. (1) IBBME/Chemical Engineering and Applied Chemistry, University of Toronto, 164 College Street, Rm 407, Toronto, ON M5S 3G9, Canada, (2) Department of Chemical Engineering, Northeastern University, 360 Huntington Ave., 342 SN, Boston, MA 02115

INTRODUCTION AND BACKGROUND

Over 80 million people suffer from cardiovascular disease, with over 36 percent of all deaths in North America today citing CVD as a cause [1]. Tissue engineering and cell injection strategies are attractive methods of regenerating or repairing ailing hearts but both research fields suffer from the limitation of selecting an ideal cell source. Recent research has identified the presence of resident cardiac progenitors in the adult heart. These rare cells represent a potential autologous, expandable source of cardiac cells if they can be isolated in high enough concentrations. To that end we propose the use of alginate-peptide hydrogels to render glass surfaces selectively adherent to various cell types in the neonatal rat heart model. In this study we modify alginate with the tetra-peptides, REDV (arg-glu-asp-val), VAPG (val-ala-pro-gly) and RGDS (arg-gly-asp-ser), which have been found to capture endothelial cells, smooth muscle cells and fibroblasts respectively [2,3]. Using this method we aim to sequentially deplete a heterogeneous heart isolate via surface adherence of cells, and enrich the cell suspension in non-adherent cardiac progenitor cells. Our goal is to characterize these novel surfaces for use in the clinical setting in the form of low-cost, easily fabricated microdevices to isolate cells from patient biopsies for further expansion.

MATERIALS AND METHODS

Alginate modification:

The peptides RGDS (American Peptides), REDV (American Peptides) and VAPG (Sigma) were attached to a low viscosity sodium salt of alginate (Sigma A2158) using EDC chemistry (EDC, S-NHS, Pierce Biotechnology). Alginate at 18.75mg/ml in MES buffer was activated by EDC for 2 hours and then reacted with the peptides at various concentrations for 20 hours. Following the 20-hour reaction with the peptides, the alginate was then dialysed for two days (3500 MWCO) and then lyophilized.

Surface modification:

12mm diameter circular glass coverslips were plasma treated and then modified alginate was immediately pipetted onto the cleaned surfaces. Coverslips were then placed into a spin coater to create a thin, uniform layer of alginate on the surfaces. The coverslips were removed from the spin coater and placed into 1mM $CaCl_2$ to gel. Gels were incubated at 4°C overnight before use.

Cell seeding:

Cells were isolated from 1-2 day old Sprague-Dawley rats as described previously [4] and a 60-minute preplating step was used to remove most of the non-adherent cardiomyocytes. The adhered preplated cells were expanded for a week and then trypsinized (a step which also kills cardiomyocytes). The alginate-modified surfaces were washed in sterile PBS containing calcium ions and preplate cells were resuspended in serum-free culture medium and carefully pipetted onto the coverslips. Unmodified alginate-coated surfaces as well as bare glass surfaces were used as controls. The seeded coverslips were incubated at 37°C for 45 minutes and then non-adherent cells were gently removed and culture medium added. Adherent cells were imaged at random regions of the surfaces and counted using image analysis software.

RESULTS AND DISCUSSION

One area of concern with this strategy is prevention of non-specific cell adhesion on the surfaces. The presence of the alginate between the exposed peptides on the hydrogel surface should prove non-adhesive to cells, allowing us to evaluate the efficacy of the peptide groups at cell capture. The total number of cells was counted and averaged over several surfaces, and the modified alginate surfaces compared to the unmodified alginate surfaces at the low end, and to the bare glass surfaces at the high end of cell adhesion.

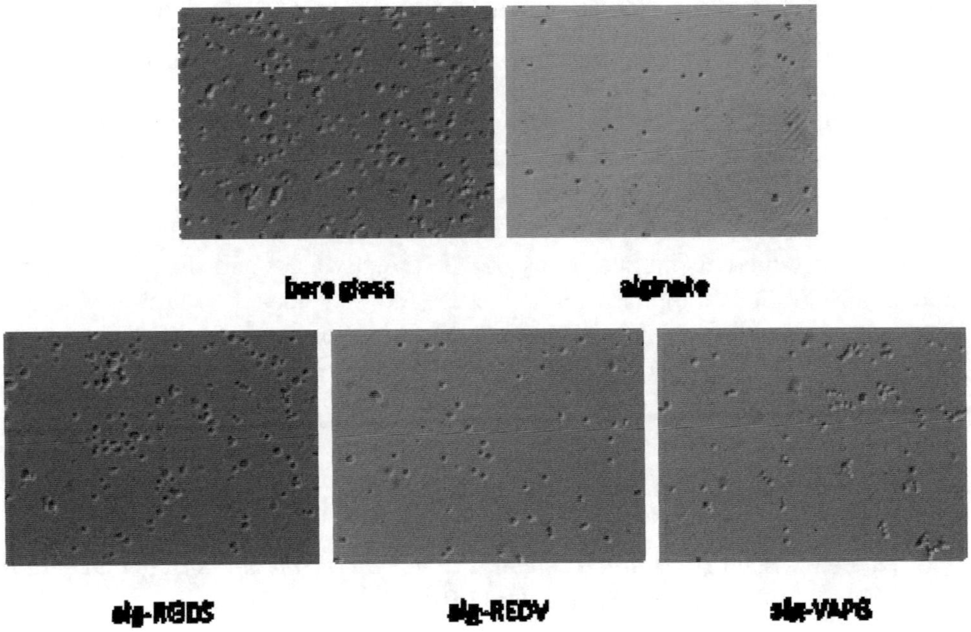

Figure 1: Brightfield images of representative regions of coated surfaces after cell seeding and incubation

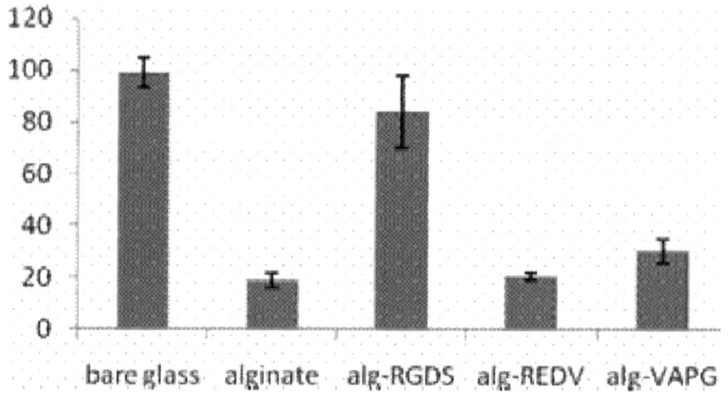

Figure 2: Cell counts of total adhered cells averaged over surface type. Peptide concentration on modified alginate surfaces is 0.6mg/ml.

At a peptide concentration of 0.6mg/ml, the RGDS surfaces showed an increased level of cell adhesion, almost comparable to bare glass levels. REDV surfaces did not show significantly greater adhesion than unmodified alginate surfaces at this concentration while VAPG surfaces showed only a slight improvement. It is to be noted that the preplate cell suspension being seeded on the surfaces is homogeneous, consisting of mainly fibroblasts, with a small percentage of endothelial and smooth muscle cells. It is to be expected then, that lower cell numbers may be seen on the surfaces specific to capturing ECs and SMCs.

A major advantage of the alginate system is that the hydrogel can be easily dissoluted using a chelating agent such as EDTA, allowing for release of any captured cells. These cells can then be cytospun onto slides and immunostained for specific endothelial, fibroblast and smooth muscle markers, such as CD31, prolyl-4-hydroxylase and alpha-smooth muscle actin respectively, to identify the captured cell types. In our ongoing studies we use these techniques to assess the effectiveness and specificity of each hydrogel at various peptide concentrations. Finally, we hope to use each surface type to serially deplete a cell suspension of ECs, SMCs and lastly FBs and characterize the cells present in the final output.

REFERENCES

1. The American Heart Association Heart Disease and Stroke Statistics -- 2008 Update

2. Plouffe, B. D., et al., *Peptide mediated selective adhesion of smooth muscle and endothelial cells in microfluidic shear flow,* Langmuir. 2007 Apr 24;23(9):5050-5.

3. Plouffe, B. D., et al., *Microfluidic depletion of endothelial cells, smooth muscle cells, and fibroblasts from heterogeneous suspensions*, Lab Chip. 2008 Mar;8(3):462-72.

4. Radisic, M. et al., *Medium perfusion enables engineering of compact and contractile cardiac tissue*, Am J Physiol Heart Circ Physiol. 2004 Feb;286(2):H507-*16*

ACKNOWLEDGEMENTS

This work was supported by grants from NSERC (DG), CFI (LOF), ORDCF (ARTEC) and University of Toronto Open Fellowships to HTHA and MFC.

136f A Microfluidic Cell Array with Individually Addressable Culture Chambers

Hsiang-Yu Wang, Chemical Engineering, Purdue University, West Lafayette, IN 47906, Ning Bao, Agricultural and Biological Engineering, Purdue University, West Lafayette, IN 47907, and Chang Lu, Agriculture and Biological Engineering, Biomedical Engineering, Chemical Engineering, Purdue University, West Lafayette, IN 47906.

Microfluidic arrays of living cells have raised a lot of interests recently due to their potential for high throughput screening of cell-based assays. This report presents a microfluidic cell array with individually addressable chambers controlled by pneumatic valves for cell culture and treatment. There are two modes for the cell array to be operated. In the first mode, different groups of cells are directed into designated chambers for culturing and observation. We demonstrate the delivery and culture of enhanced green fluorescent protein (EGFP) expressing and nonfluorescent Chinese hamster ovary (CHO) cells into specific chambers in the array. In the second mode, the chambers are first seeded with the same cell type and different reagents are delivered to specific chambers for cell treatment. We treat cells in designated chambers with Calcein AM and CellTrace calcein red-orange AM to demonstrate the principle. We envision that this microfluidic cell array technology will pave the way to automated high-throughput screening of biomolecules and drugs based on observing cellular phenotypes and responses.

136g Modelling Cell Migration In Tapered Micro-Channels

Fong Yew Leong and Keng Hwee Chiam. Large-Scale Complex Systems, Institute of High Performance Computing, 1 Fusionopolis Way, #16-16 Connexis, Singapore, 138632, Singapore

Cellular deformation in confined environment has attracted much attention in the past decade, primarily in the hopes that this can lead to efficient tools for cancer diagnostics in time to come (see review [1]). To further our understanding of cell deformability and motility, cells can be made to flow through micro-fluidic channels. These micro-fluidic devices offer the means of quantifying cell motility characteristics in-vitro through velocity measurements and cell morphology observations [1]. In this study, we are interested in modelling cell migration phenomena in micro-fluidic channels. More specifically, the immersed boundary method is implemented to track the moving cellular interface coupled with the fluid background. It is shown that the migration velocity of the cell is dramatically reduced at the inlet of the micro-channel: a phenomenon also observed experimentally. Furthermore, it is shown that the mechanical properties of the nucleus are important factors affecting cell motility and deformation in a constriction. Comparison is made between two different cases of whole cell in micro-channel flow and isolated nucleus in micro-pipette aspiration [2].

REFERENCES

[1] Suresh, S. Biomechanics and biophysics of cancer cells. Acta Biomaterialia 3, 413-438, 2007

[2] Vaziri, A. & Mofrad, M.R.K. Mechanics and deformation of the nucleus in micropipette aspiration experiment. J. Biomech. 40, 2053-2062, 2007

233a Control and Separation of Proteins In a Nanofluidic Fet Device, Using pH Gradient and Valence Charge

Youn-Jin Oh[1], Danny Bottenus[2], Cornelius F. Ivory[3], and **Sang M. Han**[1]. (1) Chemical & Nuclear Engineering, University of New Mexico, 1 University of New Mexico, MSC01 1120, Albuquerque, NM 87131, (2) Washington State University, Pullman, WA 99164, (3) Chemical & Bioengineering, Washington State University, Pullman, WA 99164

We have fabricated Si multiple internal reflection infrared waveguides embedded with a parallel array of nanofluidic channels (100 nm W ×500 nm D) and studied field-effect-transistor (FET) flow control and separation of proteins, using scanning laser confocal fluorescence microscopy (SL-CFM) and multiple internal reflection Fourier transform infrared spectroscopy (MIR-FTIRS). For fluidic FET, a DC potential is applied to a highly doped gate area in the mid-section of nanochannels, in addition to a longitudinal electric field along the nanochannels. The gate potential controls the surface charge on SiO_2 channel walls and therefore their ζ-potential. Depending on the polarity and magnitude, the gate potential can accelerate, decelerate, or reverse the flow of proteins. In addition, our MIR-FTIR analysis demonstrates that fluorescein dye molecules, used here as a pH indicator, are hydrogenated and dehydrogenated in response to the gate bias and subsequent pH shift. Using fluorescein, we have thus measured a pH shift caused by the surface charge modulation and longitudinal electrical field. We observe that this pH shift is further influenced by water electrolysis occurring at the electrodes that drive the electroosmotic flow as well as at the gate where a leakage current unavoidably flows through a thermal SiO_2 layer. Using this pH manipulation and generating a pH gradient along the nanochannels, we have conducted isoelectric focusing and separation of proteins with different isoelectric points (I*p*). In this presentation, we will further discuss protein separations, using transverse electromigration based on their different valence charges in relation to the surface charge on channel walls.

Selection of Non-Equilibrium Over-Limiting Currents: Universal Depletion Layer Formation Dynamics and Vortex Instability

Gilad Yossifon and Hsueh-Chia Chang

Department of Chemical and Biomolecular Engineering, Center for Microfluidics and Medical Diagnostics, University of Notre Dame, Notre Dame, Indiana 46556, USA

We report the first experimental verification of an electrosmotic vortex instability, first predicted by Rubinstein and Zaltzman (1988), at a micron-sized extended polarized layer on the depleted side of a homogenous nanoporous membrane, when the current through the membrane exceeds the limiting current. The vortex instability is shown to select the extent of the polarized layer and a thicker diffusion layer beyond it, and thus specify the over-limiting current. The requisite conductivity gradient across the ion-selective membrane is reproduced by a fabricated nano-microchannel junction with the former/latter having a depth smaller/larger than the Debye length. The extended polarized layer and the accompanying diffusion layer, whose much larger dimension allows fluorescent tracking, are sustained quasi-steadily with an AC field. Beyond a critical instantaneous field strength, the diffusion layer dynamics is shown to be self-similar at all scales, independent of the frequency and field strength, until the vortices arrest its growth. The vortex pair wavelength doubles in time sequentially in this self-similar region before saturation and is roughly on the order of the diffusion layer thickness.

Previous theories on ionic conduction from an electrolyte solution into an ion selective nanoporous membrane have suggested that the steady conduction flux through the membrane becomes unstable beyond the diffusion-controlled limiting current when the electrolyte concentration near the interface approaches zero (Rubinstein & Zaltzman, 2000). Unlike the Poisson-Boltzmann distribution at zero flux, which is globally stable, the polarized layer is unstable to transverse perturbations in a manner reminiscent of Rayleigh–Benard and Marangoni instabilities in thermal conduction. With any local increase in the polarized layer thickness, the diffusion layer thickness decreases and the field at the outer boundary of the polarized layer increases. This positive feedback further extends the local polarized layer thickness and the local electric field. The resulting transverse gradient in the polarized layer thickness produces a transverse field that drives a non-equilibrium and non-uniform electro-osmosis slip flow in the extended polarized layer. This vortex instability hence has its own selected length scale — the evolving diffusion layer thickness. This is in contrast to the classical electro-osmotic vortices of the 2nd time (Dukhin, 1991) around a granule, which occur because of inhomogeneous polarization due to granule curvature and exhibit length scales comparable to the granule radius (Mishchuk & Takhistov, 1995; Ben & Chang, 2002). In the theories of Rubinstein and co-workers, however, the diffusion layer thickness is pre-imposed and specified by external stirring. The implication is then the over-limiting current is stirring dependent.

Secondary instability of the steady vortices has also been predicted (Rubinstein & Zaltzman, 2000) and has been speculated to be the origin of significant noise in the measured current beyond the limiting current conditions. This mechanism is consistent with some indirect experimental observations: if the depleted diffusion layer is immobilized by a gel, a plateau is reached at saturation, and the excess electric noise disappears (Maletzki et al., 1992). However, a direct confirmation of the vortex instability of extended polarized layers and their role in determining the overlimiting current has yet to be reported.

In recent years, with the development of micro- and nano-fabrication technologies, increasing experimental evidence for this instability has been reported at the entrance of a single array of nanochannels. As in membrane nanopores, the permselectivity of a nanochannel stems from the existence of EDL overlap since nanochannels approach double layer dimensions. Effects such as concentration polarization across an array of nanochannels have been reported by several groups (e.g. Pu et al. (2004), Plecis et al. (2005) and Kim et al. (2007)). The latter group also reported electro-osmotic vortices of the 2nd time that arise from non-uniform polarization at the micro-nano junction and with dimensions comparable to the channel width. Hence, definitive and quantitative connection to true vortex instabilities with intrinsic length scales has yet to be established.

A unique universal and self-similar diffusive evolution of the field-driven diffusion layer dynamics, independent of the frequency and the instantaneous field strength, allows us to sustain and examine the vortex instability quasi-steadily with an AC field in our experiments. In the process, we confirm the predicted relationship between the instantaneous vortex pair wavelength and the evolving diffusion layer thickness (Rubinstein & Zaltzman, 2000). However, we also demonstrate from the data that, at low frequencies, the growing diffusion layer thickness saturates at an asymptotic thickness due to the vortex instability at a value that is dependent only on the field strength and the liquid mobility. This asymptotic thickness is also expected to be the one selected under DC conditions. This result has the important implication that, if hydrodynamic resistance is not too severe to suppress vortex formation (viz. too shallow a channel), the over-limiting current is independent of external stirring and is actually intrinsic to the system. We are hence able to produce the first a priori estimate of the overlimiting current that is consistent with our measured current through the nanochannel. We are also able to produce a dimensionless electro-osmotic Rayleigh number based on the quasi-diffusion layer thickness for the onset of vortices in a microfluidic slot.

The application of AC electric field is a key factor in our success to quasi-steadily sustain diffusion layers of different thickness and capture this instability. The same AC field can also suppress the instability altogether with frequencies higher than the inverse vorticity-diffusion time scale such that the macroscopic vortical flow pattern has no time to evolve. This corresponds to subcritical conditions below the critical Rayleigh number. Hence, our result has further implications on how to dynamically control (by AC forcing) the overlimiting current such that it deviates from the DC value. The Rayleigh number for the onset of vortices is dependent on the depth of the microchannel (through hydrodynamic resistance or vorticity diffusion time) and one can hence also design specific DC and AC I-V characteristics with proper adjustment of the channel geometries. Being able to control the flux through an ion-exchange membrane is both scientifically interesting because of the myriad of pertinent physics, and also important in many engineering and biological applications.

References

[1] Ben Y. & Chang H.-C., Nonlinear Smoluchowski slip velocity and micro-vortex generation, J. Fluid Mech., 461, 229-238 (2002).

[2] Dukhin S.S., Electrokinetic phenomena of the second kind and their applications, *Advances in Colloid and Interface Science*, 35, 173-196 (1991).

[3] Kim S.J., Y.C. Wang Y.C., Lee J.H., Jang H., and Han J., Concentration Polarization and Nonlinear Electrokinetic Flow near a Nanofluidic Channel, *PRL* 99, 044501 (2007).

[4] Maletzki F., Rosler H.-W. and E. Staude, Ion transfer across eelctrodialysis membranes in the overlimiting current range: stationary voltage current characteristics and current noise power spectra under different conditions of free convection, *J. Memebrane Science*, 71, 105-115 (1992).

[5] Mishchuk N.A, Takhistov P.V, Electroosmosis of the second kind, *Colloids and Surfaces A: Physicochemical and Engineering Aspects*, 95, 119-131 (1995).

[6] Plecis A., Schoch R.B., and Renaud P., Ionic Transport Phenomena in Nanofluidics: Experimental and Theoretical Study of the Exclusion-Enrichment Effect on a Chip, *Nano Lett.*, 5, 1147-1155 (2005).

[7] Pu Q., Yun J., Temkin H., and Liu S., Ion-Enrichment and Ion-Depletion Effect of Nanochannel Structures, *Nano Lett.*, 4, 1099-1103 (2004).

[8] Rubinstein I., Staude E. and Kedem O., Role of the membrane surface in concentration polarization at ion exchange membrane, *Desalination*, 69, 101 (1988).

[9] Rubinstein I. and Zaltzman B., Electro-osmotically induced convection at a permselective membrane, *Physical Review E*, 62, 2 (2000).

233c Transport of Ions and Molecules In Nanofluidic Devices

Rohit Karnik[1], Chuanhua Duan[2], Kenneth Castelino[2], Rong Fan[3], Peidong Yang[2], and Arun Majumdar[2]. (1) MIT, Cambridge, MA 02139, (2) University of California, Berkeley, (3) California Institute of Technology

Interesting transport phenomena arise when fluids are confined to nanoscale dimensions in the range of 1-100 nm. First, the length scale of electrostatic interactions in aqueous solutions becomes comparable to nanochannel size. Second, the size of the channel becomes comparable to the size of biomolecules such as proteins and DNA. Third, large surface area-to-volume ratios result in rapid rates of surface reactions compared to rates of diffusive transport. These phenomena enable us to control transport of ions and molecules in unique ways that are not possible in larger channels. Surface charge can govern ionic transport in nanochannels, enabling surface reactions and biological binding events to be sensed via their effect on nanochannel conductance. The effect of biomolecular charge is dominant at low ionic concentrations, whereas the effect of biomolecular size is dominant at higher ionic concentrations. It is also possible to locally control ionic concentrations and transport inside nanochannels through field effect in a nanofluidic transistor, which is analogous to the metal-oxide-semiconductor field effect transistor. Rapid surface reactions due to the large surface area-to-volume ratios enable a new technique of diffusion-limited patterning. This technique is useful for patterning of biomolecules and surface charge in nanochannels and can be used to fabricate a nanofluidic diode with a surface charge discontinuity for rectification of ionic transport. These experiments suggest that there is potential for controlling flows inside nanochannels for operations such as analyte focusing, pH and ionic concentration control, and biosensing.

Web Page: web.mit.edu/karnik/www/home.htm

233d Analytes Preconcentration and Separation in Nanofluidic Channels

Thomas C. Gamble[1], Yi Zhang[2], Zhen Yuan[3], Alexander Neumann[4], Gabriel P. Lopez[5], Steven R. J. Brueck[6], and **Dimiter N. Petsev**[1]. (1) Chemical & Nuclear Engineering, University of New Mexico, 1 University of New Mexico, MSC01 1120, Albuquerque, NM 87131, (2) Department of Chemical & Nuclear Engineering, UNM, 209 Farris Engineering Center, Albuquerque, NM 87131, (3) Department of Chemical Engineering, University of New Mexico, Advanced Materials Laboratory, 1001 University Blvd. SE, Suite 100, Albuquerque, NM 87106, (4) Center for High Technology Materials, UNM, 1313 Goddard SE, Albuquerque, NM 87106, (5) Chemical and Nuclear Engineering, University of New Mexico, Farris Engineering Center MSC01 1120, 1 The University of New Mexico, Albuquerque, NM 87131, (6) University of New Mexico, UNM, 1313 Goddard SE Rm 145, Albuquerque, NM 87106

Controlling the field distribution in fluidic channels allows to controllably manipulating dissolved analytes. Examples include the field gradient methods for focusing and preconcentration of charged molecules. In this study we suggest a method for shaping the electric field in nanofluidic channels that are fabricated on a Si wafer. The method exploits the conductivity of SiO_2 that allows precise shaping of the externally applied field in the fluid. We present a theoretical model of this system that is based on Maxwell electrodynamics equations. The model has been experimentally verified by conducting an analyte focusing experiment in Si/SiO_2 nanaofluidic channels (see the Figure below). Nanofluidic channels can have dimensions comparable to the electric double layer thickness. We analysed the double layer effect on the transport of electric current, analyte electrophoresis and electroosmotic flow. We have shown that very narrow channels exhibit unique features (e.g., ion selecvtivity) that disappear with increasing the cross-sectional dimension.

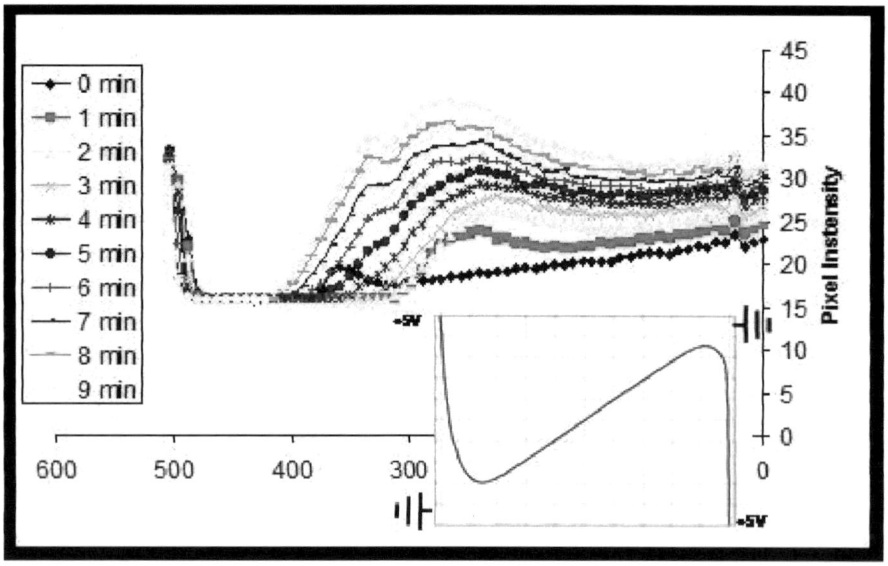

233e Diffusivity Effects In Charged Particle Transport In Nanochannels

David A. Boy, Maria Napoli, Frederic Gibou, Igor Mezic, and Sumita Pennathur. Mechanical Engineering, UCSB, Santa Barbara, CA 93106

Fluid dynamic behavior in nanochannels is dominated by the interactions between the channel surface and the fluid. An example of a particularly significant surface-fluid interaction involves the electric double layer. The electric double layer is a region of high charge density close to the channel surfaces, where counter-ions are attracted to the surface charge and co-ions are repelled. Electric double layer thicknesses are on the order of nanometers and can therefore significantly affect fluid flow at the nanoscale.

In this work, we develop a continuum model for the transport of charged, finite-size particles in electroosmotically driven flow that includes the effects of finite-thickness electric double layers. In our model, we assume a dilute solution of particles and a DC electric field applied along the axis of the channel. Under these assumptions, previous work has shown that the combination of cross-channel gradients in velocity and electric potential allows for a separation mechanism, where higher counter-ion charged analytes have faster velocities [1]. The model we propose extends this previous analysis, incorporating Taylor-Aris dispersion theory. By theoretical means and direct numerical simulation, we show that differences in particle size, and therefore in diffusivities, can enhance, diminish, or even reverse the separation behavior compared to the case of particles that have the same size but different charge.

To validate our model, we performed separation experiments on simple-cross, isotropically-etched, 5μ m wide by 1μ m deep fluidic channels. We injected buffered flow (10mM borate buffer, pH 9.2) seeded with charged polystyrene spherical particles (50 and 100nm diameter, fluorescently labeled) into a channel, and measured fluorescence intensity 15mm downstream from the injection point. Preliminary results showed that separation of charged particles in the nanochannels qualitatively matches our theory.

[1] "Free-solution oligonucleotide separation in nanoscale channels", Pennathur, S., Baldessari, F., Santiago, J., et al. Analytical Chemistry, 79, 8316-8322, (2007).

233f Electromigration Current Rectification In a Single Cylindrical Nanopore

Jung Yeul Jung, Mechanical Engineering, Arizona State University, 1711 S Rural Rd ECG G346 Mechical and Aerospace Eng, Tempe, AZ 85287-6106, Trevor J. Thornton, Electrical Engineering, Arizona State University, 1711 S Rural Rd ECG G346 Mechical and Aerospace Eng, Tempe, AZ 85287-6106, and **Jonathan D. Posner**, Mechanical & Aerospace Eng. & Chemical Engineering, Arizona State University, 1711 S Rural Rd ECG G346 Mechical and Aerospace Eng, Tempe, AZ 85287-6106.

We present the experimental measurements of electromigration current through a single cylindrical nanopores. A cylindrical nanopore with ~175 nm diameter was fabricated using standard clean-room semiconductor processing techniques. The fabricated nanopore exhibits electromigration rectification with low concentration solutions (100 μ M KCl) and nearly symmetric conductance with in stronger ionic strength solutions (100 mM KCl) . Our results show that the rectification is consistent with the two distinct electromigration regimes: (1) Ohmic regime where the electric double layers are thin compared to the pore size ($\kappa a>1$) and (2) thick or overlapping electric double layers ($\kappa a<1$) where the conductance of the nanopore is strongly influenced by the surface conductance in the EDL. Thick EDL's ($\kappa a<1$) exhibit current rectification due to concentration polarization while thinner EDLs show nearly symmetric conductance.

302a Nanofluidic Systems for Concentration and Detection of Biomolecules

Jongyoon Han, Biological Engineering / Electrical Engineering and Computer Science, MIT, 36-841, 77 Mass Ave., Cambridge, MA 01730

Recent developments in fabrication technology allow one to reliably build nanofluidic channels and filters with critical dimension around ~10nm, fully integrated with microfluidic channel systems in an arbitrary manner. These nanochannels act as perm-selective membrane such as Nafion(R), which are widely used in diverse applications. Such nanochannels can be used in lieu of polymeric membrane, much much higher mechanical stability. In addition, experiments done in such nanochannels can provide important insights about complex membrane and interface phenomena. In this talk, several recent examples of nanofluidic engineering will be presented, including the nanofluidic enhancement of biomolecule binding kinetics. One of the bottlenecks of the current bioassay is the diffusive transport limitation (between the target molecule and surface-mount antibody) at low target molecule concentrations. Nanofluidic channels can be used to enhance surface binding reactions, since the target molecules are confined to the surfaces which are coated with specific binding partners. Moreover, diffusion-limited binding can be significantly enhanced if the molecules are steered into the nanochannels via either pressure-driven or electrokinetic flow. By monitoring the nanochannel impedance, which is sensitive to surface binding, low analyte concentrations have been detected electrically in nanofluidic channels within response times of 1-2 hours. This represents a ~54 fold reduction in the response time using convective flow compared to diffusion-limited binding. At high flow velocities the presented method of reaction kinetics enhancement is potentially limited by force-induced dissociations of the receptor-ligand bonds. Optimization of this scheme could be useful for label-free, electrical detection of biomolecule binding reactions within nanochannels on a chip.

302b Nanoscale Physics of Induced-Charge Electrokinetic Phenomena

Martin Z. Bazant, Massachusetts Institute of Technology, 77 Mass. Ave., 2-363B, Cambridge, MA 02139, Mustafa Sabri Kilic, Mathematics, Massachusetts Institute of Technology, Cambridge, MA 02139, Brian D. Storey, Franklin W. Olin College of Engineering, Needham, MA 02492, and Armand Ajdari, Physico-Chimie Theorique, ESPCI, 10 rue Vauquelin, Paris, France.

The current theory of electrokinetics assumes a dilute solution of point-like ions in chemical equilibrium with a surface whose double-layer voltage is of order the thermal voltage, $kT/e = 25$ mV. In nonlinear electrokinetic phenomena, such as AC electro-osmosis and induced-charge electrophoresis, several Volts $\approx 100\, kT/e$ are applied to the double layer, so the theory breaks down and cannot explain many observed features. In this paper, we review the relevant literature and argue that, under such a large voltage, counterions condense near the surface, even for dilute bulk solutions, and the classical concepts of ``compact layer" and ``shear plane" must be generalized. Using simple continuum models, we predict (at least) two basic trends at large voltages: (i) the growth of a condensed layer decreases the double-layer differential capacitance, and (ii) viscosity increase with counterion condensation reduces the electro-osmotic mobility. The former may explain observed flow reversal in AC electro-osmotic pumps, while the latter may explain the decay of all induce-charge electro-osmotic phenomena at high salt concentration, although a complete theory is still lacking. Through several colloidal examples, such as the electrophoresis of an uncharged metal sphere in an asymmetric electrolyte, we predict that induced electro-osmotic flows are generally ion-specific at large voltages, even in the absence of Faradaic reactions.

Web Page: math.mit.edu/~bazant/ICEO

302c Discrimination of DNA Structure Using Single Protein Channels Supported on Glass Nanopore Membranes

Ryuji Kawano, Hao Sun, Anna Schibel, Chris Cauley, and Henry White. Chemistry, Univerity of Utah, 315 S 1400 E, Salt Lake City, UT 84112

Abstract. Ion channel recordings using transmembrane protein channel and synthetic nanopores are being applied as single-molecule biosensors and for electrical readout of polynucleotide sequences. Alpha-hemolysin (a-HL), a protein toxin isolated from Staphylococcus aureus has been extensively employed for as a biological pore for these purposes. a-HL self-assembles from monomer peptide in artificial lipid bilayers to form a heptamer nano-scale ion channel across the lipid membrane. Ion channel recordings using a-HL are typically performed using a planer bilayer that spans a ~50 um diameter orifice in a polymer-based membrane such as Teflon. The stability of the bilayer in this configuration is generally limited to several hours. We recently reported the fabrication of a single nanopore embedded in a glass or quartz membrane, and a method of preparing a suspended bilayer across the nanopore orifice. The glass nanopore (GNP) membrane is a single conical shaped nanopore embedded in a ~50 um-thick glass membrane. GNP membranes are nearly ideal supports for ion channel recordings due to the greatly improved stability of the lipid bilayer over the nanopore orifice, and electrical properties (low leakage current and low capacitance). Continuous ion channel recordings with the GNP of over 400 hours, and at potential biases of up to ±0.8 V have been demonstrated [1]. The GNP support is suitable for stochastic detection of single molecules and analysis of DNA structure [2]. We will present a description of the GNP membrane preparation and applications of this methodology in biosensing and analysis.

References

(1) White, R.; Ervin, E.; Yang, T.; Chen, X.; Daniel, S.; Cremer, P.; White, H. J. Am. Chem. Soc. 2007, 129, 11766-11775.

(2) Ervin, E.; Kawano, R.; White, R.; White, H. Anal. Chem. 2008, 80, 2069-2076.

302d Experimentally and Theoretically Observed Native pH Shifts In a Nanochannel Array

Danny Bottenus[1], Youn-Jin Oh[2], Sang M. Han[2], and Cornelius F. Ivory[3]. (1) Washington State University, Pullman, WA 99164, (2) Chemical & Nuclear Engineering, University of New Mexico, 1 University of New Mexico, MSC01 1120, Albuquerque, NM 87131, (3) Chemical & Bioengineering, Washington State University, Pullman, WA 99164

Lab on a chip (LOC) technology provides a powerful platform for simultaneous separation, purification, and identification of low concentration multicomponent mixtures. As the characteristic dimension of LOC devices decreases down to the nanoscale, the possibility of containing an entire lab on a single chip is becoming a reality. This research examines one of the unique physical characteristics of nanochannels, in which native pH shifts occur. As a result of the electrical double layer taking up a significant portion of a 100 nm wide nanochannel, electroneutrality no longer exists in the channel causing a radial pH gradient.

This work describes experimentally observed pH shifts as a function of ionic strength using the fluorescent pH indicator 5-(and-6)-carboxy SNARF®-1 and compares it to a model developed using Comsol Multiphysics. At low ionic strengths (~ 3 mM) the mean pH shift is approximately 1 pH unit whereas at high ionic strengths (~ 150 mM) the mean pH shift is reduced to 0.1 pH units. An independent analysis using fluorescein pH indicator is also presented supporting these findings.

Two independent non-linear simulations coupling the Nernst-Planck equation describing transport in ionic solutions subjected to an electric field and Poisson's equation to describe the electric field as it relates to the charge distribution are solved using a finite element solver. In addition, the effects of chemical activities are considered in the simulations. The first numerical simulation is based on a surface ζ-potential which underestimates the experimental results by about a factor of two to four for most ionic strengths. A modified model assuming that SNARF and fluorescein molecules are able to diffuse into the hydrolyzed SiO2 phase, and in the case of the SNARF molecule, able to bind to the active silanol groups agrees quantitatively with experimental results.

302e Stretching Genes

Winston Timp[1], Utkur Mirsaidov[2], and Gregory Timp[2]. (1) Medicine, Johns Hopkins University, 720 Rutland Ave., Ross 1064, Baltimore, MD 21205, (2) University of Illinois at Urbana-Champaign, 405 N. Mathews Ave., 3041 Beckman Institute, Urbana, IL 61801

To identify and discriminate alleles, Affymetrix currently uses a hybridization technique in conjunction with microarrays on high-throughput platforms to measure the difference in thermal stability between matched and mismatched target-probe pairs of dsDNA. We plan to use a related strategy for detecting SNPs targets that leverages the stretching transition. Our strategy employs short (10-12 nucleotide long) probes hybridized to the target cognate sites on ssDNA and uses high-speed SMFS to measure the threshold voltage associated with stretching or unzipping the templates in a pore <2.5nm in diameter. The transition from dsDNA to ssDNA, called the helix-coil transition, is a key element of this strategy. SMFS has been used to analyze the helix-coil transition, revealing a dichotomy in the force required to dissociate base pairs: the force is different depending on whether the DNA is unzipped by pulling parallel to the bases, or stretched by pulling transverse to the base-pairs. About F_{unzip} = 10 - 30 pN of force is required to unzip dsDNA. On the other hand, when dsDNA is stretched beyond its contour length, it undergoes a cooperative stretching transition near F_{str} = 60 - 70 pN. This overstretching transition has been interpreted as force-induced melting in which the two DNA strands break apart and unwind. Synthetic nanopores can be used to accomplish both measurements because large forces can be applied using the field in the pore, and because the pore diameter can be controlled with sub-nanometer precision.

We studied the helix-coil transition by measuring the permeability of hairpin DNA (hpDNA) through a nanopore in a silicon nitride membrane. Nucleic acid hairpins or stem-loops, formed from self-complementary sequences, are found regularly in DNA and RNA secondary structure. The dynamics of hairpin structure has already been carefully scrutinized with SMFS accomplished using α-HL. The structure is not static; there is a folded (closed) conformation consisting of a single-stranded coil, a double-stranded stem and a coil, and an unfolded (open or melted) state. The folded (closed) conformation is characterized by a low enthalpy due to base pairing in the stem, while the open state has high entropy due to the large number of configurations.available to ssDNA. Nanopore SMFS is superior to other methods because it can done without a molecular tether or anchor, and the loading rate or the force can be held constant and range up to 100Vnsec, while maintaining high throughput (~1000 molecules/second). But unlike α-HL, a synthetic nanopore affords us the opportunity to constrain the transverse motion of the molecule through control of the diameter.

We systematically investigated hpDNA permeability through a membrane as a function of the applied voltage and pore diameter. To determine if hpDNA permeates the membrane through the pore, we analyzed the extract taken from the anode using qPCR. We observe a threshold voltage for translocation of the hairpin through the pore that depends sensitively on the diameter. We assume that the threshold voltage corresponds to the minimum force required to impel a hairpin through the pore. Assuming that the stem frustrates the permeation of hpDNA, a large force corresponding to the change in free energy, ΔG, of the helix-coil transition will be required to dissociate the bases and induce the translocation of DNA.

We have found a dependence of the threshold voltage on the pore diameter; the threshold voltage collapses from about ~2V to ~0.5V as the pore size decreases from 1.5nm to 1nm, indicative of a dramatic change (x4) in the force required to induce the transition. Assuming that a pore with $d<1.5$nm excludes dsDNA and precludes stretching, we attribute this change to the difference between stretching and unzipping DNA. Ostensibly, the force required to dissociate base-pairs is different depending on whether the DNA is unzipped by pulling parallel to the bases or stretched by pulling transverse to the base-pairs. It takes less force to unzip DNA than to stretch the backbone. With the hpDNA's conformation unconstrained by the pore walls, the bases can rotate so that much of the force is applied along the axis of the hydrogen bonds connecting the bases. And so the helix-coil transition occurs by unzipping. However, if the hairpin penetrates deep within the pore, only a small component of the force is directed along the axis of the hydrogen bonds.

In summary, we observe a threshold voltage for translocation of the hairpin through the pore that depends sensitively on the diameter and the secondary structure of the DNA. For a diameter $1.5<d<2.3$nm the threshold corresponds to the force required to stretch the stem of the hairpin, while for $1.0<d<1.5$ nm, the threshold collapses because the stem unzips with a lower force than required for stretching. In related work, we have observed a threshold voltage for the rupture of the bond between a restriction enzyme and DNA that can be used to discriminate single nucleotide polymorphisms. Although speculative, it seems likely that the threshold for a hairpin to permeate the pore is related to the free energy and molecular stability, but an unambiguous interpretation requires knowledge of the molecular configuration in the pore: i.e. whether the molecule enters oriented coil- or loop-first. This information could be recovered through force spectroscopy studies of the translocation of hairpins one at a time through the pore, but first we have to sort out the relationship between the current transients and the configuration of the molecule in the pore.

333a Modification of Cells Using a High-Throughput Microelectroporator

Daniel Stark[1], Yoonsu Choi[2], Sourindra Maiti[3], Dean Lee[3], Robert Raphael[4], Thomas Killian[1], Laurence Cooper[3], and Sibani Lisa Biswal[5]. (1) Physics and Astronomy, Rice University, 1900 Rice Blvd, MS-61, Dell Butcher Hall, Room 100, Physics and Astronomy, Rice University, Houston, TX 77005, (2) Pediatrics Research, M.D. Anderson Cancer Center, 7455 Fannin St., Houston, 77054, (3) Pediatric Research, M.D. Anderson Cancer Center, 7455 Fannin St, Houston, 77054, (4) Bioengineering, Rice University, 1900 Rice Blvd, MS-61, Dell Butcher Hall, Room 100, Physics and Astronomy, Rice University, Houston, TX 77005, (5) Chemical Engineering, Rice University, 1900 Rice Blvd, MS-61, Dell Butcher Hall, Room 100, Physics and Astronomy, Rice University, Houston, TX 77005

Genetic modification of cells to redirect their specificity towards desired antigens expressed on target cells is a promising therapy for the treatment of diseases including cancer. However, conventional viral and especially non-viral gene transfer methods for gene transfer are laborious, inefficient, and rely on full-scale good manufacturing practice GMP (GMP) facilities. There is a current trend to produce micro and nano-size devices that can perform physical, chemical, and biological processes on a small scale with the same efficiency as conventional macroscopic systems. We have designed a microfluidic/ electrotransfer system for continuous insertion of transgenes coded by RNA and/or DNA into cells via electroporation. Electroporation is an indispensable and safe method for delivery of drugs and genes into cells and has been used in clinical trials1,2. Our device is capable of handling a large number of cells, on the order of those needed clinically for therapy. Further, it can operate in a large parameter space, optimizing frequency, amplitude, waveform shape, and number of pulses experienced. By optimizing the electroporation parameters, external mRNA or DNA will be incorporated into a cell, enabling the cell to produce its own drugs. Finally, the components of the device are adaptable to be implemented in a clinical setting. Poration occurs when the transmembrane potential across a cellular membrane is greater than a threshold potential. The membrane potential is approximately given by Ed, where E is the electric field and d is the membrane thickness3. Once the membrane has porated, mRNA or DNA passes into the cell and the desired function is expressed. Our device was fabricated using standard photolithographic techniques, shown in Figure 1.Jurkat cells at a concentration of about 106 cells/mL were flowed through our device at a rate of 2 μ L/s. Propidium iodide (PI), a nucleic acid stain, at a 1:10 dye/media volume ratio was used to stain cells to determine viability and poration after exposure. In Figure 2, we show how flow through the device and exposure to PI affects cell viability. Compared to the control population, we see no significant cell death (less than 10%) on the time scale required for electroporation. This demonstrates that tens of minutes of exposure to PI or to high sheer stress does not cause cellular death, and that a clinically relevant number of cells pass successfully through the device. For experiments showing electroporation, we used an optimization procedure described by Canatella and Prausnitz in order to determine the electric pulse amplitude, duration, and number of pulses each cell should be is exposed to for optimal electroporation4. We selected 10 millisecond square wave pulses at 100 Hz and a field of 833 V/cm to porate Jurkat cells. The flow velocity sets the number of pulses each cell sees to 12. Figure 3 shows a collection of cells that have passed through the electrode and are fluorescing, indicating that PI has entered the cell. Our next experiments are to check cell viability after electroporation. We have shown a clinically relevant number of cells may pass through our electroporator without cell death and are modified successfully. REFERENCES: 1Costa et al. A method for genetic modification of human embryonic stem cells using electroporation. Nature protocols (2007) vol. 2 (4) pp. 792-6 2Van Tendeloo et al. Highly efficient gene delivery by mRNA electroporation in human hematopoietic cells: superiority to lipofection and passive pulsing of mRNA and to electroporation of plasmid cDNA for tumor antigen loading of dendritic cells. Blood (2001) vol. 98 (1) pp. 49-56 3Krassowska et al. Modeling Electroporation in a Single Cell. Biophysical Journal (2007) vol. 92 (2) pp. 404-417 4Canatella et al. Prediction and optimization of gene transfection and drug delivery by electroporation. Gene therapy (2001) vol. 8 (19) pp. 1464-9

333b Rapid Fabrication and Characterization of 3-D Branched Microvascular Flow Networks

Jen-Huang Huang, Jeongyun Kim, Arul Jayaraman, and **Victor M. Ugaz**. Artie McFerrin Department of Chemical Engineering, Texas A&M University, 3122 TAMU, College Station, TX 77843

Standard photolithography-based micromachining techniques are widely used to construct 2D microchannel networks, but the inherently planar nature of these processes limits their usefulness in the creation of 3D structures. More recent developments have enabled fully 3D flow networks to be produced using processes including solid freeform fabrication, stereolithography, and 3D printing. But many of these methods involve serial 'direct writing' processes that require timescales on the order of hours to days and are not practical for mass production. In addition, no single technique has proven ideal to construct microchannel networks that incorporate a wide range of size scales (μm to mm). Here, we describe a new process that enables 3-D branched microvascular networks to be constructed in a single step. This technique employs an electrostatic discharge phenomenon that occurs when a dielectric medium is energized by a strong electric field and subsequently discharged to form branched "tree-like" channels in polymer substrates. Suitable space charge distributions are generated by irradiating the sample with an electron beam so that the energy released upon discharge is sufficient to locally vaporize and fracture the material, leaving behind a network of branched channels in a tree-like fractal structure. Beam intensity and spatial irradiation profile (width and penetration depth) can be adjusted to control the location and morphology of the discharge structures.

The embedded patterns exhibit a self-similar fractal network structure, with channel characteristic dimensions ranging from approximately 100 nm to 1 mm in diameter. The formation of interconnected networks in the discharge structure was demonstrated by nucleating discharges at multiple sites along the substrate perimeter. Interconnectivity of the network between the different nucleation sites was demonstrated by injecting an aqueous dye into the network. The diameter of the microchannels was estimated to be ~ 20 microns using fluorescent beads and confocal microscopy. A microfluidic analog of the discharge structure was developed in parallel to investigate the shear stress and pressure drop in narrow channels in order to determine the feasibility of culturing cells inside these networks. Bovine aortic endothelial cells (BAEC) cultured in the microfluidic analog were viable for 72 h at flow rates of ~ 1 μ L/min and demonstrate that cell culture within channels of similar dimensions in discharge structures can be carried out. We have also used this process to form microchannel networks in poly (lactic acid) substrates, suggesting that this technique could be used to generate vascular flow structures in substrates relevant for tissue engineering applications.

333c Negative Selection Separation of Cells In Microfluidic Devices

James V. Green[1], Brian D. Plouffe[1], Milica Radisic[2], and **Shashi K. Murthy**[3]. (1) Chemical Engineering, Northeastern University, 360 Huntington Ave., 006 Mugar, Boston, MA 02115, (2) IBBME/Chemical Engineering and Applied Chemistry, University of Toronto, 164 College Street, Room 407, Toronto, ON M5S 3G9, Canada, (3) Department of Chemical Engineering, Northeastern University, 360 Huntington Ave., 342 SN, Boston, MA 02115

Microfluidic cell separation systems have emerged as attractive alternatives to traditional techniques in recent years. These systems offer the advantages of being able to handle small sample volumes and at the same time achieve highly selective separation at low cost. Microfluidic devices with surface-immobilized adhesion molecules can achieve separation by exploiting ligand-receptor interactions. When ligands are immobilized on the microfluidic channel surfaces, the resulting cell capture devices offer the typical advantages associated with microfluidic systems, with the added benefit of not requiring complex fabrication schemes or pre-processing incubation. This presentation will describe how a microfluidic system of devices coated with peptides can be utilized to deplete a cell suspension of endothelial cells, smooth muscle cells, and fibroblasts in order to isolate a fourth cell type by negative selection. The peptides utilized are arg-glu-asp-val (REDV), val-ala-pro-gly (VAPG), and arg-gly-asp-ser (RGDS). Separation is achieved by continuous flow and by the provision of large surface areas for cell attachment. The significance of this approach is that it can be utilized to isolate stem and progenitor cell populations from digested tissue as a precursor to conventional tissue engineering on scaffolds or cell-based regenerative therapeutics. Furthermore, this approach could be an effective way to isolate stem/progenitor cells whose markers are not fully characterized.

Microfluidic cell electroporation using a mechanical vavle

Jun Wang, M. Jane Stine and Chang Lu[a)]
Department of Agricultural and Biological Engineering, School of Chemical Engineering, Birck Nanotechnology Center, Bindley Bioscience Center, Purdue University, West Lafayette

Introduction

Electroporation is a technique that applies electric field to transiently permeabilize the cell membrane to deliver impermeant molecules into cells. Due to its wide applications to gene/drug delivery, electroporation has been adapted in microfluidic devices by fabricating microscale electrodes and structures that apply short electric pulses [1-3]. The fabrication of microscale electrodes presents a difficulty for lowering the cost of these microfluidic chips. Furthermore, an electric pulse generator is needed to generate pulses with durations ranging from microseconds to milliseconds. The pulse generator also adds to the cost and complexity of the overall system. In recent work, microfluidic channels with geometric variation were applied to electroporation and electrofusion of flowing cells using a common dc power supply [4, 5]. While the technique doesn't require a pulse generator and microfabricated electrodes, it cannot be applied to *in situ* electroporation of adherent cells. The adhesion to the substrate can be important for cellular functions of adherent cell types. Thus electroporating cells while they adhere to the substrate is often desired due to the minimum disturbance to the cell culture.

Here we demonstrate a microfluidic technique that electroporates both suspended cells and cells adhering to the bottom of a microfluidic channel using a common dc power supply, taking advantage of a fast-response microscale mechanical valve [6]. The mechanical valve turns on/off a dc electric field established in a microfluidic channel rapidly by physically connecting/separating the ionic buffer. By virtue of that, we are able to effectively produce electric pulses of milliseconds using this approach. We investigate the effects of the valve dimensions, the actuation pressure, and the voltage applied on the performance of the valve in terms of pulse generation. We also examine the viability of cells after electroporation using the device and demonstrate the electropermeabilization of cells by an impermeant dye SYTOX green.

Experiment

1. Microchip fabrication

The design of the device is shown in Figure. 1. The device was fabricated using multilayer soft lithography with substantial modifications[6]. The control layer master (photoresist/3 inch silicon wafer) was made using a negative photoresist SU-8 2025 (Microchem) with a thickness of 33 μm. The fluidic layer master was made using a positive photoresist AZ 9260 (Clariant) with a thickness of 25 μm. The fluidic layer master was baked at 120 ℃ for 2 min to generate a rounded cross sectio n for the channel. The thickness of the photoresist was translated into the depth of the microfluidic channel. Both control and fluidic layers of the device were molded using PDMS of the same composition (GE Silicones RTV 615, MG Chemicals, mass ratio of A:B=10:1). The fluidic layer had a thickness of 133 μm (formed by spinning of the liquid PDMS prepolymer at 1200 rpm for 30s) which left the thickness of the PDMS membrane between the fluidic channel and the control channel around 108 μm. The control layer had a thickness ~0.5 cm. The two layers were bonded together

[a)]Author to whom correspondence should be addressed; email: changlu@purdue.edu

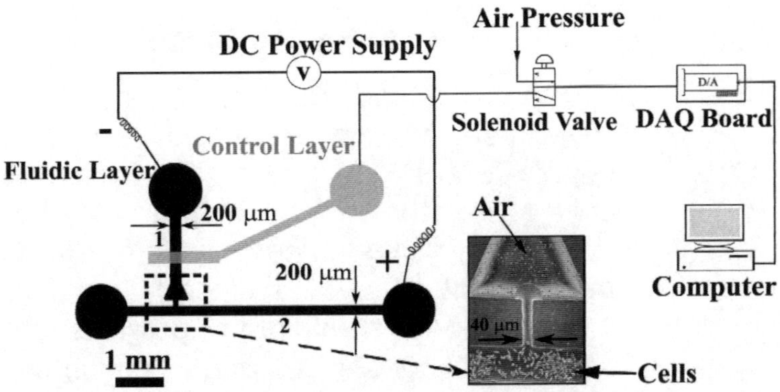

Figure 1. The design of the microfluidic electroporation device. The control channel (grey) is on top of the fluidic channel (black) with a PDMS membrane in between. The valve locates in the vertical fluidic channel 1 and the cell culture is in the horizontal fluidic channel 2. A common dc power supply is kept on during the operation of the device.

upon contact after oxidizing the two PDMS surfaces using a Tesla coil (Kimble/Kontes) in atmosphere and the combined layers were then bonded to a clean glass slide using the same oxidation method. Finally the entire device was baked in an oven at 80 ℃ for 2 hr. The PDMS device was exposed to UV light in a biological safety cabinet overnight for sterilization before use. The fluidic channels were also silanized by exposing to (Tridecafluoro-1,1,2,2-tetrahydrooctyl)trichlorosilane vapor for 2 hr in order to increase the insulation property of the valve when closed.

2. Cell Sample Preparation and Seeding in the Microfluidic Device.

Chinese hamster ovary (CHO-K1) were cultured in plastic tissue culture flasks at 37℃, under 5% CO_2 in the Dulbecco's modified Eagle's medium (DMEM, Mediatech Inc.) supplemented with 10% (v/v) fetal bovine serum (FBS, Sigma), penicillin (100 units/ml, Sigma), and streptomycin (100μg/ml, Sigma). Cells were diluted at a ratio of 1:5 every 2 days to maintain them in the exponential growth phase (~1×10^6 cells/ml). They were harvested and resuspended in the culture medium at a concentration of 1×10^8 cells/ml for seeding in the microfluidic device. Before seeding, the horizontal channel 2 (Figure 1) was pre-coated with 1 mg/ml fibronectin (Sigma) (incubation for 0.5 hr at 80℃[7]), before flowing CHO cells into channel 2. The cells were cultured in the channel 2 for 12 hr in a 37℃, 5% CO_2 incubator while the culture medium slowly infused through the channel 2 under gravity, before the experiment unless otherwise noted. All the operations involving seeding and culture were done exclusively in channel 2 by keeping air in the channel 1. The cell viability was examined by flowing 1μM SYTOX green into the channels at 1 hr after electroporation.

3. Fluorescence Microscopy.

The microfluidic device was mounted on an inverted fluorescence microscope (IX-71, Olympus). The epifluorescence excitation was provided by a 100W mercury lamp, together with brightfield illumination. The excitation and emission from cells peameabilized with DNA dye SYTOX green (Molecular Probes) were filtered by a fluorescence filter cube (exciter HQ480/40, emitter HQ535/50, and beamsplitter Q505lp, Chroma technology Corp.).

4. Microchip Operation.

The layout and the setup of the microchip are shown in Figure 1. The valve was actuated by pressurizing the control channel (filled with water) using a fast-response solenoid valve (ASCO Scientific). The pressure pushed down the PDMS membrane to contact the glass bottom and closed the fluidic channel. The solenoid valve was controlled by a LabVIEW program which designated the time of the valve being opened.

After the cell culture, phosphate buffer (8 mM Na_2HPO_4, 2 mM KH_2PO_4, and 250mM sucrose, pH=7.2) was used to replace the culture medium and fill in the entire fluidic channels (both channels 1 and 2) before electroporation. A dc power supply (1AA12P30, Ultravolt, with current monitoring) provided constant dc voltage when the chip was in operation.

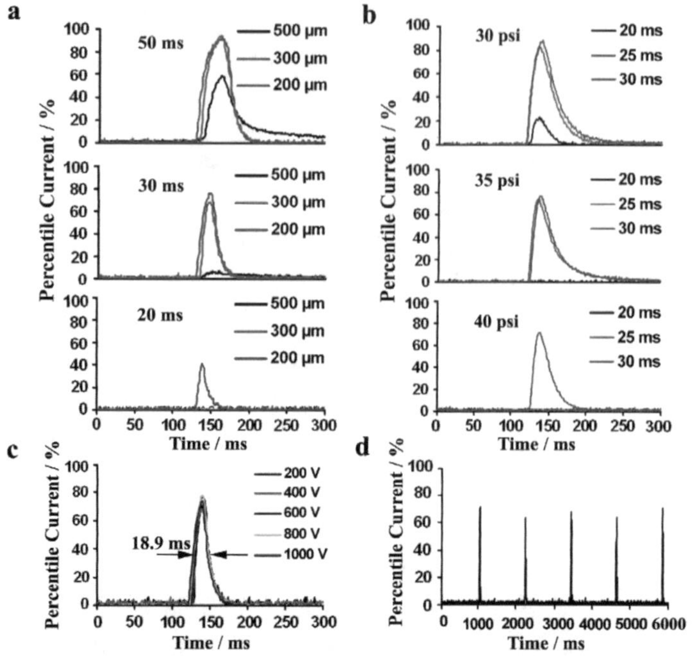

Figure 2. The effects of the control channel width, the actuation pressure, and the applied voltage. **(a)** The percentile current-time curves with different control channel widths and valve opening times. The actuation pressure is 40 psi and the voltage is 400 V. **(b)** The percentile current-time curves with different actuation pressure and valve opening times. The control channel width is 300 µm and the voltage is 400 V. **(c)** The percentile current-time curves with different voltages provided by the power supply. The control channel width is 300 µm, the valve opens for 30 ms, and the actuation pressure is 40 psi. **(d)** A series of pulses generated by opening the valve for 30 ms within an interval of 1.2 s for 5 consecutive times. The control channel width is 300 µm, the voltage is 400 V, and the actuation pressure is 40 psi.

Results and Discussion

We used a simple multilayer PDMS device and a common dc power supply to carry out the electroporation of cells (as shown in Figure 1). The multilayer valve effectively creates an open-circuit while it is closed due to its insulating nature. While the power supply provided constant and continuous voltage, the valve was operated to have a close-open-close sequence to generate a pulse which electroporated the cells in the channel 2. It is found that long time soaking of the PDMS valve with the culture medium could weaken its electric insulating ability and increase the background current significantly. As shown in the inset image in Figure 1, the triangle shape structure was used to keep the culture medium out of channel 1 during the whole cell seeding process, taking advantage of the hydrophobicity of PDMS. The cells were exclusively cultured in channel 2.

We examined the effects of the width of the control channel, the actuation pressure, and the dc voltage across the channel on the generation of the pulse. The current was monitored between the two electrodes while the valve was operated to close, then open for a designated period (milliseconds), and close again. To present the current in percentile, we designated the current with the valve fully open as 100% and that with two electrodes connected to an insulator as 0 under a certain voltage. In Figure 2a, the overall voltage applied was 400 V and the valve was actuated by 40 psi pressure. The width of the control channel was varied from 200 to 500 µm. As the width of the control channel increases, the response of the valve gets more retarded. A 500 µm valve failed to generate a pulse when the valve was actuated to open for less than 30 ms by the computer. On the other hand, a 200 µm valve seems to response most promptly to the changing pressure. It generated a pulse as short as 20 ms and always gave the highest percentile current at its peak. The high peak percentile current indicates that the 200 µm valve was closest to be fully open during the operation compared with the other valves with wider control channels. We conducted tests with varied pressure and voltages using devices with the control channel of the width of 300 µm. In Figure 2b, the effect of the actuation pressure is shown. Higher pressure produced less "tail" for the pulse. However, it also became harder to generate short pulses (e.g. with 40 psi no pulses were produced when the duration was set at 20 and 25 ms.). The percentile current profiles generated by the same device were fairly reproducible at different voltages from 200 V to 1000 V (Figure 2c). The average full width at half maximum (FWHM) of the pulses was 18.9 ms with a standard deviation of 1.1 ms when the valve opened for 30 ms. In Figure 2d, a series of pulses was generated by opening the valve for 30 ms within an interval of 1.2 s for 5 consecutive times. The relative standard deviation (RSD) in the peak height was 5.7% and the RSD in the FWHM was 1.2%. The uniformity of a pulse series is better when the valve opens for longer periods. We were able to model the field intensity in the device using software package COMSOL 3.2 when the valve was entirely open and the effects of cells were ignored. The local field intensity is proportional to the electrical current based on Ohm's law. Therefore we could calculate the field intensity variation experienced by the cells in channel 2 based on the percentile current data in Figure 2. For example, using the data in Figure 2b, it is found that the peak intensity of the pulse generated by 30 psi was 327 V/cm with the valve opening for 30 ms, while it was 309 V/cm and 87 V/cm with the valve opening for 25 and 20 ms, respectively. The current data indicate that the actuation of the valve is effective in generating a pulse with reproducible duration and magnitude.

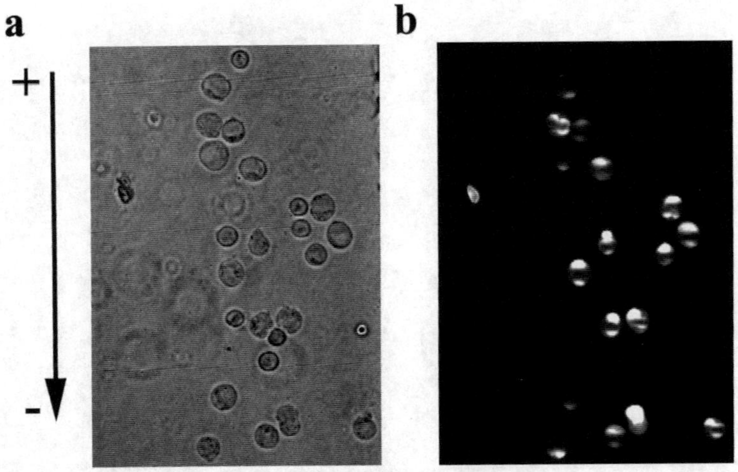

Figure 3. Electropermeabilization of suspended CHO cells (before culturing) with a pulse of 30 ms and a pulse intensity of 209 V/cm. The channel contains 5 μM SYTOX green during the electroporation. The images were taken 5 min after the electroporation. **(a)** Phase contrast. **(b)** Fluorescence.

The devices with the control channel width of 300 μm were chosen for CHO cell electroporation tests. All these cell electroporation experiments were done by having the valve open once for 30 ms with an actuation pressure of 40 psi. SYTOX green was the model molecule we used for testing the electropermeabilization. It is a DNA counterstain that is impermeant to live cells and becomes green-fluorescent when bound to DNA. Figure 3 shows that the suspended cells were stained by SYTOX green at 5 min after a single pulse. The cells mostly had their two opposite poles stained that were aligned with the field direction. Similar results were obtained using electroporation based on a pulse generator and are characteristic of electroporative delivery [8, 9]. Figure 4a shows the time sequenced images of electropermeabilization of adherent CHO cells after applying one pulse using the valve at time 0. The images reveal that the delivery occurred first at the cathode-facing hemispheres of the cells. Such asymmetric transport during electroporation has been described before [8]. For the electroporation of adherent CHO cells, we further examined the viability of cells (the percentage of live cells) and the SYTOX green permeabilization rate among live cells after the electroporation under various pulse intensities (peak field intensities experienced by the cells). When the viability was measured, we carried out the electroporation and then flowed SYTOX green into the channel 1 hr later. The stained cells were the dead ones (with leaky membrane) in this case. When the permeabilization rate was measured, we first obtained the percentage of cells being stained by having 1μM SYTOX green in the channel during the electroporation.

The percentage of stained cells included both dead cells and live/electropermeabilized ones in this case. By deducting the percentage of dead cells based on the viability data, we obtained the permeabilization rate that represented the live and permeabilized cells among the starting cell population. The optimal permeabilization rate was reached at the pulse intensity of 279 V/cm with 51% of the cell population permeabilized and alive. Electroporation in this experiment occurred at field intensities lower than those commonly reported in the literature for CHO cells (~400 V/cm)[5, 10]. This is possibly related to the long pulse duration (the electroporation threshold is a slowly decreasing function of the pulse duration) and the shape of the pulse.

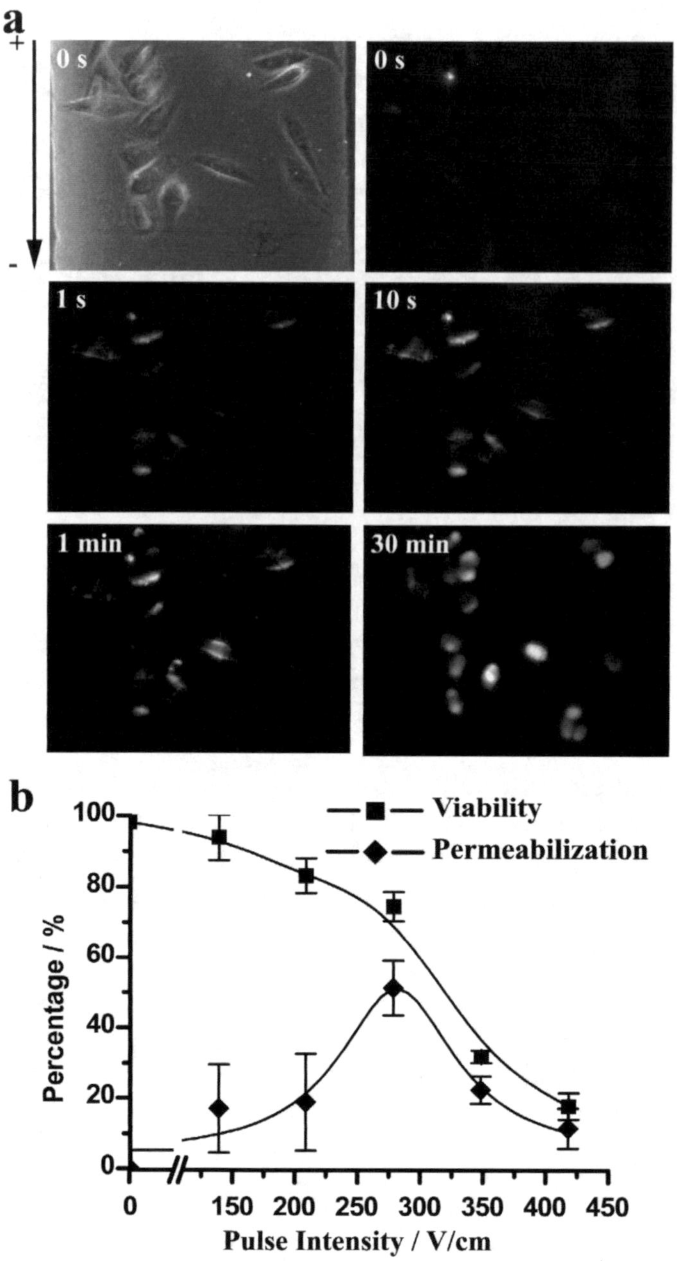

Figure 4. (a) Electropermeabilization of adherent CHO cells with a pulse of 30 ms and a pulse intensity of 349 V/cm. The pulse is applied at time 0. **(b)** The viability and permeabilization rate of adherent CHO cells at different pulse peak intensities. The viability represents the percentage of live cells among the starting cell population. The permeabilization rate represents the percentage of both live and permeabilized cells among the starting cell population.

Conclusion

We demonstrated that a simple microfluidic device could effectively produce cell fusion under continuous DC voltage, when coupled with chemical linking method. This technique offers efficiency comparable to that of conventional electrofusion technique which requires expensive and complicated apparatus. Cell fusion was conducted at single pair level in the microfluidic channel and the physical dimensions of the channel effectively prevented more than 3 cells from fusing together. These features are important when cell fusion needs to be conducted based on scarce cell sources such as primary cells. This demonstration will help establish microfluidics as a viable platform for studying a number of important biological problems involving cell fusion. Further studies are needed to characterize the properties of fused cells produced by this technique.

Acknowledgement

This research was supported by a Purdue Research Foundation fellowship and Purdue University.

References

1 Lin, Y. C., Li, M. and Wu, C. C. (2004) Simulation and experimental demonstration of the electric field assisted electroporation microchip for in vitro gene delivery enhancement. *Lab on a Chip* **4**, 104-108

2 Yamauchi, F., Kato, K. and Iwata, H. (2005) Layer-by-layer assembly of poly(ethyleneimine) and plasmid DNA onto transparent indium-tin oxide electrodes for temporally and spatially specific gene transfer. *Langmuir* **21**, 8360-8367

3 Huang, Y. and Rubinsky, B. (2003) Flow-through micro-electroporation chip for high efficiency single-cell genetic manipulation. *Sensors and Actuators a-Physical* **104**, 205-212

4 Wang, J. and Lu, C. (2006) Microfluidic cell fusion under continuous direct current voltage. *Applied Physics Letters* **89**, -

5 Wang, H. Y. and Lu, C. (2006) Electroporation of mammalian cells in a microfluidic channel with geometric variation. *Analytical Chemistry* **78**, 5158-5164

6 Unger, M. A., Chou, H. P., Thorsen, T., Scherer, A. and Quake, S. R. (2000) Monolithic microfabricated valves and pumps by multilayer soft lithography. *Science* **288**, 113-116

7 Gu, W., Zhu, X. Y., Futai, N., Cho, B. S. and Takayama, S. (2004) Computerized microfluidic cell culture using elastomeric channels and Braille displays. *Proceedings of the National Academy of Sciences of the United States of America* **101**, 15861-15866

8 Tekle, E., Astumian, R. D. and Chock, P. B. (1994) Selective and Asymmetric Molecular-Transport across Electroporated Cell-Membranes. *Proceedings of the National Academy of Sciences of the United States of America* **91**, 11512-11516

9 Golzio, M., Teissie, J. and Rols, M. P. (2002) Direct visualization at the single-cell level of electrically mediated gene delivery. *Proceedings of the National Academy of Sciences of the United States of America* **99**, 1292-1297

10 Golzio, M., Mora, M. P., Raynaud, C., Delteil, C., Teissie, J. and Rols, M. P. (1998) Control by osmotic pressure of voltage-induced permeabilization and gene transfer in mammalian cells. *Biophysical Journal* **74**, 3015-3022

333e Microfluidic Control of Spatial Nutrient Gradients for Plant Culture on a Chip

Satya Gowthami Achanta[1], David Henthorn[1], and Chang-Soo Kim[2]. (1) Department of Chemical and Biological Engineering, Missouri University of Science and Technology, 143 Schrenk Hall, Rolla, MO 65409, (2) Dept.of Electrical and Computer Engineering, Dept. of Biological Sciences, Missouri University of Science and Technology, Rolla, MO 65409

Of paramount importance in the culture of cells and organisms is efficient access to various nutrients. A useful tool in the study of organism response to various nutrients is a microfluidic cell where precise spatial control of species concentration is readily available. In this work, we have fabricated devices which create linear gradients of nutrients, such as oxygen or potassium and nitrate ions, directly in the cell culture chamber. All microfluidic structures were fabricated from the negative tone photoresist SU-8 because of its high optical clarity and for its compatibility with on-chip optical waveguide manufacture. Nutrient-sensitive fluorophores were immobilized in conjuction with optical waveguide posts to allow for characterization of the cell culture chamber microenvironment.

333f Complex 3D Scaffolds for Cell Culture

Noy Bassik, George Stern, Mustapha Jamal, and David H. Gracias. Department of Chemical and Biomolecular Engineering, Johns Hopkins University, 125 Maryland Hall, 3400 N Charles Street, Baltimore, MD 21218

Tissues often have complex toplogies, but it is still challenging to construct microstructured scaffolds for culturing cells in non-planar 3D geometries. We describe a strategy to self-assemble complex micropatterned structures such as coils, spirals and cylinders. We investigated the use of these substrates as scaffolds for cell culture in 3D.

333g Microfluidic Neuromuscular Control of Insect Micro-Air-Vehicles

Aram J. Chung and David Erickson. Sibley School of Mechanical and Aerospace Engineering, Cornell University, Ithaca, NY 14853

This work describes the intimate fusion of microsystems and physiology though the implantation of a microfluidic device into living insects. Specifically, we focus on the control of Manduca Sexta moth flight activity using an implanted, low power electrokinetic drug delivery device. This forms one of the critical components in our development of "Insect-Micro Air Vehicles (I-MAVs)" which aim to fuse nanodevice technology with living systems. The goal of this system is to provide "on-command" chemically induced immobilization and subsequent reanimation of the otherwise autonomous insect. In this paper, we demonstrate the chemical control over the level of insect locomotor activity by releasing different neurotransmitters (i.e. L-Glutamic Acid, L-Aspartate Acid, Gamma-aminobutyric acid, Betta-Alaine) into wing muscle. We also provide results of our adult survivability data for pupae stage implanted microdevices. Our direct toxin injection experiments will also be presented along with results from comprehensive study of a low power electroactive drug delivery system.

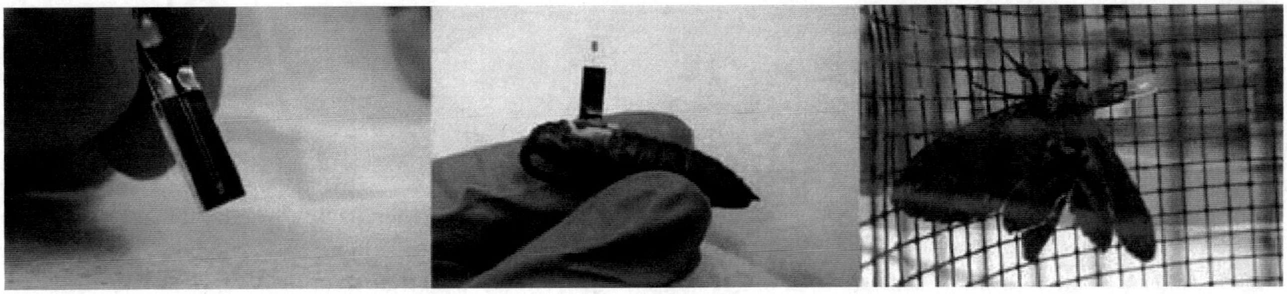

393a Quantitative Proteomic Analysis of Membrane Proteins Based on Gel and Shotgun Techniques

Trong Khoa Pham and Phillip C. Wright. Chemical and Process Engineering, The University of Sheffield, CHELSI, The Sheffield Bioincubator, 40 LeavyGreave Road, Sheffield, S3 7RD, United Kingdom

Most membrane proteins play essential cellular functions. However, from a proteomics aspect, the analysis of membrane proteins is still challenging due to the hydrophobic nature of membrane proteins that makes it difficult for proteomic analysis, especially for quantitative proteomics. Until now, proteomic analysis of membrane proteins has mostly been based on 2-DE followed by mass spectrometry. However, this technique is limited by the difficulty of solubilizing and resolving membrane proteins on gels, furthermore, identification of proteins based on gels is often tedious. Unlike 2-DE techniques, shotgun proteomics offers a useful alternative to gels. Although, recently, many different shotgun methods have been used to analyze enriched membrane fractions using detergents, organic solvents, and additives, in the present of trypsin or other proteolytic agents, none of these methods has been used for quantitative membrane proteomic analysis yet.

Sulfolobus solfataricus is a member of archaeal domain, that has been received much attention since the mechanisms relating to information processes represent a simpler version of eukaryotic equivalents. *S. solfataricus* is a thermoacidophilic crenarchaeon that grows optimally at conditions of 80°C and pH 3. To survive under these extreme conditions at high temperature and low pH, *S. solfataricus* needs to optimise its biological behaviour, especially at the bioenergetic aspect. How it manages to do this is not yet clear. Additionally, most bioenergetic processes, especially oxidative processes, are performed via membrane proteins, but the numbers of these proteins that have been experimentally identified as well as quantified are still low. Therefore, development of techniques to overcome this issue is necessary.

We describe a method for extracting *S. solfataricus* membrane proteins, as well as applying both techniques based on 2-DE and shotgun workflows for quantitative proteomic of *S. solfataricus* membranes. Here, for the first time, the application of isobaric tags for relative and absolute quantitation (iTRAQ) for quantitative analysis of membrane proteins is performed. A comparison between 2-DE and iTRAQ techniques for quantitative proteomics of membranes is also discussed. Moreover, the application of a new MS instrumentation based of an ion trap with electron transfer dissociation (ETD) is also investigated in comparison with Qq-TOF instruments employing collision induced dissociation (CID).

393b High-Throughput Metalloproteomics: X-Ray Fluorescence Imaging Paired with Electrophoresis

Lydia A. Finney[1], Yasmin Chishti[2], Tripti Khare[2], Carol Giometti[2], and Stefan Vogt[3]. (1) X-ray Science and Biosciences Divisions, Argonne National Laboratory, 9700 S Cass Ave, Argonne, IL 60439, (2) Biosciences Division, Argonne National Laboratory, 9700 S Cass Ave, Argonne, IL 60439, (3) X-ray Science Division, Argonne National Laboratory, 9700 S Cass Ave, Argonne, IL 60439

A critical need exists for technology that pairs with electrophoresis to identify and measure the metal ion component of metalloproteins in complex mixtures. These metals serve numerous roles as structural nuclei, catalytic cofactors, and in signaling. Their directed availability to proteins, a process itself tightly regulated, is a means of cellular control of protein activity. Yet, while gel electrophoresis allows us to readily capture information about proteins themselves, the metals they bind often elude comparable analysis. This work describes a simple, comprehensive approach for identifying and quantifying the metal ion component of metalloproteins in complex samples using native-PAGE and synchrotron x-ray fluorescence imaging. The approach is equally applicable to both 1D and 2D electrophoresis. Both the identification and quantitation of each metal bound to a protein spot is demonstrated, and the potential of this technique is illustrated in initial applications.

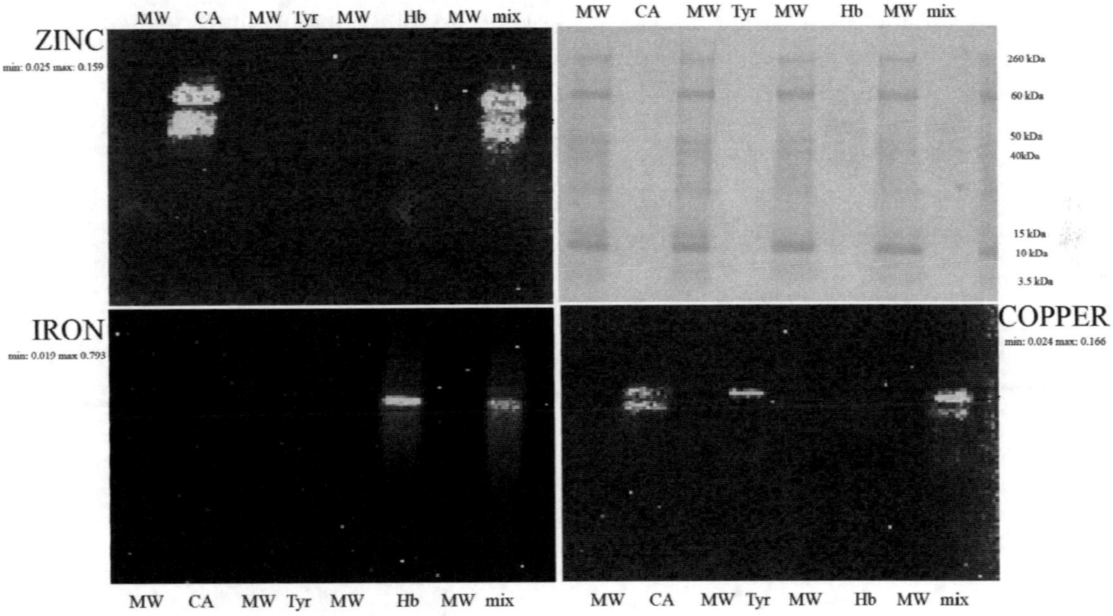

* Note: Markers are denatured. Native protein migration is affected by mass, charge, and shape.

393c 2D Gel Western Blotting Using Antibodies against Acetyl, Sumo and Ubiquinin Groups on Proteins Is a Useful Tool for Studying Disease Processes

Nancy Kendrick, Matthew Hoelter, and Jon Johansen. Kendrick Labs Inc, 1202 Ann St, Madison, WI 53713

Understanding protein interactions within a cell is key to unraveling disease processes. Genomic studies alone will not suffice. Many methods exist for resolving proteins mixtures including high throughput HPLC and capillary electrophoresis, and low throughput SDS PAGE, and 2D electrophoresis (2DE). Different proteins have dramatically different characteristics imparted by rigid tertiary structure dependent on amino acid order, so no one method suffices to resolve all the proteins within cells. Of the known methods, 2DE has the highest resolution by far for resolving protein mixtures, with the capability of separating several thousand proteins on a single gel. But although the proteins are resolved, they cannot be easily visualized and identified. General staining methods including Coomassie blue, fluorescence and silver have notoriously poor sensitivity.

Western blotting may be the solution. This method of immunostaining is very sensitive, on the order of 10-100 fold higher than any staining method depending on the antibody, and can delve very deep into the proteome. While most antibodies are generated against specific proteins, new classes of antibodies are now being developed against post-translational modifications (PTM) of proteins such as phosphorylation and glycosylation. Subsets of proteins with specific PTMs affected by disease processes can be found by 2DE Western blotting and identified by mass spectrometry. In this presentation we will report results of Western blotting experiments using anti-acetyl, anti-ubiquitin, and anti-SUMO antibodies. Validation results will be presented and sensitivity of the method described along with problems and applications. Western blotting with new antibodies against acetyl, ubiquitin, and SUMO groups should be an important method for studying protein turnover within cells.

Web Page: www.kendricklabs.com

393d Fluorescence Visualization of Newly Synthesized Proteins In Mammalian Cells

Julie C. Liu[1], Kimberly E. Beatty[2], Fang Xie[3], Daniela C. Dieterich[4], Erin M. Schuman[4], Qian Wang[3], and David A. Tirrell[2]. (1) School of Chemical Engineering, Purdue University, 480 Stadium Mall Drive, West Lafayette, IN 47907, (2) Division of Chemistry and Chemical Engineering, California Institute of Technology, 1200 E. California Blvd., Pasadena, CA 91125, (3) Department of Chemistry and Biochemistry, University of South Carolina, 631 Sumter St., Columbia, SC 29208, (4) Howard Hughes Medical Institute and Division of Biology, California Institute of Technology, 1200 E. California Blvd., Pasadena, CA 91125

Modern proteomic methods enable efficient identification of proteins present in whole cells or in isolated organelles, but a thorough understanding of the proteome requires insight into protein localization as well as protein identity. Previous work demonstrated visualization of newly synthesized proteins in bacterial cells through co-translational introduction of an alkynyl amino acid followed by selective Cu(I)-catalyzed ligation of the alkynyl side chain to the fluorogenic dye 3-azido-7-hydroxycoumarin. In this work, homopropargylglycine (Hpg), a methionine (Met) analog, was used to selectively tag newly synthesized mammalian proteins in a method similar to conventional pulse-chase labeling, and labeled proteins were visualized via fluorescence. Pulses as short as 15 minutes with no chase provided a five-fold fluorescent enhancement compared to controls. We demonstrate that the extent of labeling can be controlled through the ratio of Hpg to Met and the concentration of the copper (I) catalyst. This fluorescent-labeling method worked on a wide variety of cell types, and, in many cases, the most intense fluorescence appeared to localize to nucleolar structures. This study thus demonstrates a facile method for visualizing proteins based on the timing of protein synthesis even when the sequence, structure, or function of the proteins is unknown.

393e Nuclear Proteomics

Miguel Barthelery, Ugur Salli, and Kent Vrana. Pharmacology, Penn State College of Medicine, 500 university drive R130, Hershey, PA 17033

Cellular responses originate in the nucleus where gene expression is initiated. The genetic material and its expression are managed and maintained by the nuclear proteome. We postulated that expression proteomic technologies, applied to nuclear protein extracts, would enrich our understanding of the molecular reprogramming that occurs during in vitro differentiation of human embryonic stem cells. However, this approach is challenged with the over-representation of histones and the presence of nucleic acids, both of which significantly interfere with isoelectric focusing. We have developed an effective and reproducible technique to deplete nuclear protein extracts of histones and nucleic acids in a single step. This method significantly improves the efficacy and the resolution of 2-DIGE separation, enhancing the quality of image analysis for nuclear protein extracts.

PROTEOMICS-LEVEL IDENTIFICATION OF KINETICALLY STABLE PROTEINS BY DIAGONAL 2D SDS-PAGE

Ke Xia, Songjie Zhang, and Wilfredo Colón, Rensselaer Polytechnic Institute, Troy, NY

Introduction

Kinetic stability is a poorly understood property of a select group of naturally occurring proteins that are kinetically trapped in their native conformations (*1*). Kinetic stability can be best explained by illustrating the unfolding process as a simple equilibrium reaction between two protein conformations, the native folded state (N) and the unfolded state (U), separated by a transition state (TS) (Fig. 1). Since the height of the TS free energy determines the rate of folding and unfolding, kinetically stable proteins possess a TS with an unusually high free energy, resulting in extremely slow unfolding rates that virtually trap the protein in its native state (Fig 1A). Even when the overall change in Gibbs free energy (ΔG) may be favorable for unfolding under extreme solvent conditions, such as high concentrations of denaturant, the high activation energy of the TS significantly slows down the unfolding rate. It has been suggested that the presence of a high energy barrier separating the folded and unfolded states is an evolutionary feature to allow proteins to maintain activity in the extreme conditions they may encounter *in vivo* (*1*). This is consistent with the observation that thermodynamic stability alone does not fully protect proteins that are susceptible to irreversible denaturation and aggregation arising from partially denatured states that become transiently populated under physiological conditions (*2*). Therefore, the development of a high-energy barrier to unfolding may serve to protect susceptible proteins against such harmful conformational "side-effects". The physical basis for kinetic stability is poorly understood and no structural consensus has been found to explain this phenomenon. It appears that no single structural feature exists to explain kinetic stability, and perhaps this property may be achieved by different means, depending on the individual protein.

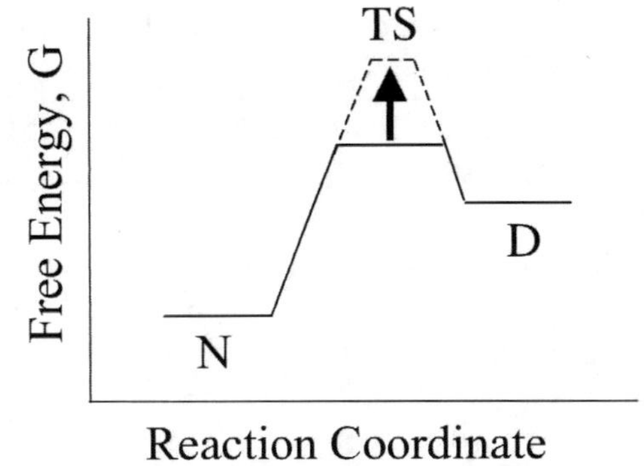

Fig. 1. Free energy diagram to illustrate the higher unfolding energy barrier for a kinetically stable protein under native conditions, as compared to that of a normal protein (represented by the dash line).

How to Analyze Protein Kinetic Stability?

The concept of kinetic stability as an alternative explanation for protein stability, independent from thermodynamic stability, was introduced in the early 90's (*3, 4*). Since then, kinetic stability has been shown to be an important physical property of some proteins (*5-7*). Since kinetically stable proteins are characterized by having a large energetic barrier that trap them in their native state (Fig. 1), determining the protein unfolding half-life under native conditions provides a quantitative description of kinetic stability. These experiments require pure protein and a fluorescence spectrometer/stopped-flow instrument to measure the rate of protein unfolding at different concentrations of denaturant (e.g. urea). The log of the unfolding rate is plotted against the denaturant concentration and the extrapolation of the linear plot to 0M denaturant yields the unfolding rate under native conditions. Thus, the small percentage of proteins that are kinetically stable combined with the lack of a simple method to detect and quantify kinetic stability has limited the study and understanding of this important property of proteins.

SDS Resistance: A New and Simple Method to Screen For Kinetic Stability

Manning and Colón discovered that kinetically stable proteins are resistant to denaturation by the detergent sodium dodecyl sulfate (SDS) (*8*) and this has led to the development of a simple boiled-unboil assay (Fig. 2) for detecting kinetic stability in proteins (*8*). Recently, we developed and optimized a diagonal 2-dimensional (D2D) SDS-PAGE version of the assay that preserves the key unboil-boil step to allow the screening of complex mixtures of proteins, such as cell lysates (*9*). Only microgram amounts of sample are needed, and the method is suitable for identifying kinetically stable proteins present in cell extracts without need for purification, thus making it a fast and inexpensive method for the detection of kinetic stability in proteins.

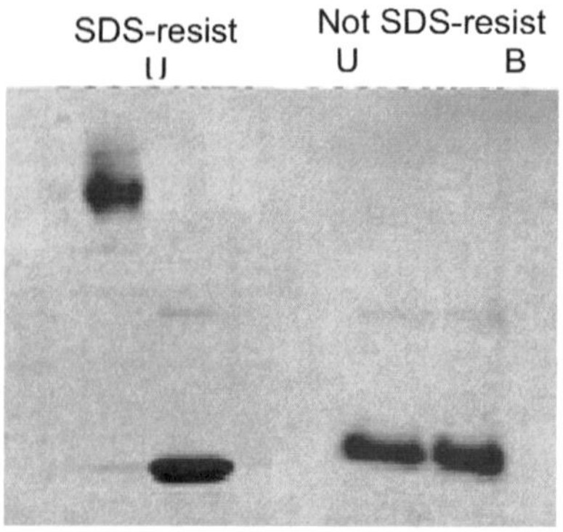

Fig. 2. 1D SDS-PAGE assay for identifying kinetically stable proteins. The protein is incubated in sample buffer (1% (w/v) SDS), divided in two, and one of them is boiled. The migration on the gel of the unboiled (U) and boiled (B) samples are compared. The "U" sample of kinetically stable proteins will migrate less than the "B" sample, while for SDS-labile proteins, both samples will have the same migration.

The concept behind our D2D SDS-PAGE method is illustrated in Fig. 3. Most proteins are easily denatured by SDS even without heating, and thus migrate the same distance in both dimensions of the gel analysis, resulting in a diagonal line of spots across the gel. However, SDS-resistant proteins travel a shorter distance in the 1st dimension due to minimal SDS binding. In the 2nd dimension, the SDS-resistant proteins become denatured by the heating step and exhibit the expected migration on SDS-PAGE, and consequently are located off the diagonal. We applied our D2D SDS-PAGE method to the cell lysate of *E. coli* and, as expected, the 2D gel showed a strong diagonal indicating that most proteins in *E. coli* are not kinetically stable (*9*). Proteomics analysis involving LC-MS/MS led to the identification of 50 non-redundant proteins that were SDS resistant (i.e. kinetically stable). Structural and functional analyses of a subset (44) of these proteins with known 3D structure revealed some potential structural and functional biases towards and against kinetic stability. This simple D2D SDS-PAGE assay should allow the widespread investigation of kinetic stability, including the proteomics-level identification of kinetically stable proteins in different systems, potentially leading to a better understanding of the biological and pathological significance of this intriguing property of proteins.

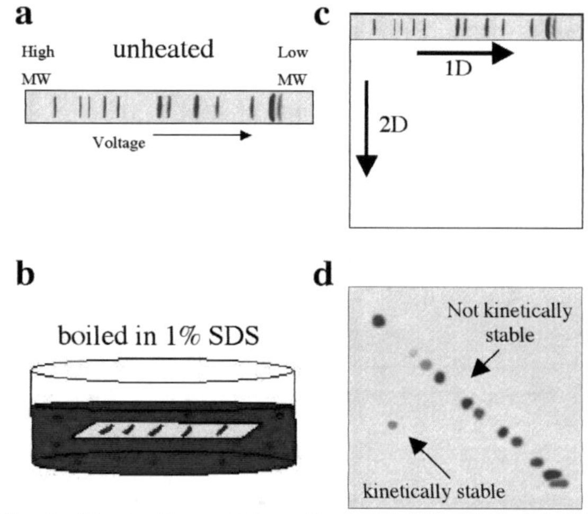

Fig 3. Illustration of the diagonal 2D SDS-PAGE procedure. **a)** The 1st dimension SDS-PAGE separation is performed. **b)** The lane is excised and incubated in boiling sample buffer containing 2% SDS. **c)** The proteins on the gel strip are transferred onto a larger gel and a 2nd dimension SDS-PAGE separation is performed. **d)** The non-kinetically stable proteins migrate equal distances in both SDS-PAGE dimensions to form a diagonal pattern, while kinetically stable proteins will migrate less in the 1st dimension, giving rise to spots that will fall below the gel diagonal. Figure is from reference 9.

References

1. Cunningham, E. L., Jaswal, S. S., Sohl, J. L., and Agard, D. A. (1999) Kinetic stability as a mechanism for protease longevity, *Proc. Natl. Acad. Sci. U S A 96*, 11008-11014.
2. Plaza del Pino, I. M., Ibarra-Molero, B., and Sanchez-Ruiz, J. M. (2000) Lower kinetic limit to protein thermal stability: a proposal regarding protein stability in vivo and its relation with misfolding diseases, *Proteins 40*, 58-70.
3. Baker, D., and Agard, D. A. (1994) Kinetics versus thermodynamics in protein folding, *Biochemistry 33*, 7505-7509.
4. Baker, D., Sohl, J. L., and Agard, D. A. (1992) A protein folding reaction under kinetic control, *Nature 356*, 263-265.
5. Hammarstrom, P., Jiang, X., Hurshman, A. R., Powers, E. T., and Kelly, J. W. (2002) Sequence-dependent denaturation energetics: A major determinant in amyloid disease diversity, *Proc. Natl. Acad. Sci. U S A 99 Suppl 4*, 16427-16432.
6. Hammarstrom, P., Wiseman, R. L., Powers, E. T., and Kelly, J. W. (2003) Prevention of transthyretin amyloid disease by changing protein misfolding energetics, *Science 299*, 713-716.
7. Lynch, S. M., Boswell, S. A., and Colon, W. (2004) Kinetic stability of Cu/Zn superoxide dismutase is dependent on its metal ligands: implications for ALS, *Biochemistry 4 3*, 16525-16531.
8. Manning, M., and Colón, W. (2004) Structural basis of protein kinetic stability: Resistance to sodium dodecyl sulfate suggests a central role for rigidity and a bias towards beta sheet structure, *Biochemistry 43*, 11248-11254.
9. Xia, K., Manning, M., Hesham, H., Bystroff, C., Lin, Q., and Colón, W. (2007) Identifying the subproteome of kinetically stable proteins via diagonal 2D SDS-PAGE. , *Proc. Natl. Acad. Sci. 104*, 17329-17334.

393g A Novel Mixed-Integer Linear Optimization Framework for the Identification of Post-Translationally Modified Proteins Using Etd/ecd Tandem Mass Spectrometry

Peter A. DiMaggio Jr.[1], Richard Baliban[2], Benjamin A. Garcia[3], and Christodoulos A. Floudas[1]. (1) Department of Chemical Engineering, Princeton University, Dept of Chemical Engineering; A215 Engineering Quadrangle, Princeton, NJ 08544, (2) Chemical Engineering, Princeton University, Dept of Chemical Engineering; A215 Engineering Quadrangle, Princeton, NJ 08544, (3) Department of Molecular Biology, Princeton University, Dept of Chemical Engineering; A215 Engineering Quadrangle, Princeton, NJ 08544

Accurate identification of post translational modifications (PTMs) can give insight into the dynamic proteome and illuminate the role and function of specific proteins in vivo. Current approaches for the identification of protein modifications utilize a "bottom up" approach, where the proteins are enzymatically digested into smaller peptides that are subsequently ionized and fragmented using Collision Activated Dissociation (CAD) to derive their sequence information [1]. The identification of all the modifications present in a protein sequence hinges on the successful identification of the PTM modifications of its corresponding peptides. This protocol can be limited by (a) insufficient elution and detection of the all the peptides that cover the entire sequence of the protein and (b) false identifications at the peptide level. Additional complications arise when using CAD to study labile modifications such as phosphorylation, glycosylation, or sulfonation. In these instances, the preferred reaction is often the cleavage of the PTM as opposed to the backbone of the peptide, resulting in a high intensity peak corresponding to the parent mass less the cleaved modification. The advent of Electron Capture Dissociation (ECD) [2-3] and Electron Transfer Dissociation (ETD) [4-7] have enabled researchers to address the aforementioned issues associated with CAD in a complementary fashion.

ECD and ETD both involve the reaction of an electron with a highly protonated cation to form an odd-electron peptide. This process induces large amounts of backbone cleavage to yield c- and z-ions that are analogous to the b- and y-ions produced from CAD. Unlike CAD, ECD and ETD cleavage is very weakly affected by the composition and number of amino acids in the peptide and provides more fragmentation coverage than CAD alone. Both ECD and ETD also prevent cleavage of labile modifications and these PTMs are fully present on the c- and z-ions produced during cleavage. The aforementioned benefits make ECD and ETD well-suited for the "top-down" analysis of post-translationally modified proteins. A further benefit of ECD/ETD analysis is the quantitative information obtained from the spectral data. A full protein sequence is eluted into the spectrophotometer, so it is expected that the resulting ratios of peaks intensities will correspond quite closely to concentration ratios of the ions. If two proteins with slightly different modifications eluted at the same time, measurement of the peak intensity ratios can give a good approximation of the ratio of protein present in the sample.

Both ECD and ETD were used to extensively study the modifications on histone H3 [8]. The variety of modifications present on the N-terminus of the protein represents a "histone code" that must be fully characterized to understand the cellular processes in which H3 is involved. To fully appreciate the depth of potential modification combinations for histone H3, hydrophilic interaction chromatography (HILIC) was used in combination with gas-phase isolation of species with different parent masses. HILIC chromatography will separate the different H3 proteins mainly by number of acetyl groups and secondly by degree of methylation. Such a separation will allow for better validation of the spectral data by comparing the potential modifications with the elution time. Consequent analysis has revealed potential relationships between K14, K18, and K23 acetylations as well as interplay between K4 methylation and the previously mentioned acetylations.

In this work, we will present de novo methods based on mixed-integer linear programming (MILP) for the identification of post-translationally modified peptides using ETD and ECD tandem mass spectra, with a particular emphasis on the study of histone H3. Though a vast amount of de novo and database methods have been developed to analyze unmodified spectra [9-11], the ETD/ECD can illuminate certain in vivo features such as internally cleaved sequences and spliced sequences that cannot be determined once the original sample is digested with trypsin. To our knowledge, the only currently available software for the analysis of ETD spectra is the Open Mass Spectrometry Search Algorithm [12], which is a database method. We will present a de novo mixed-integer linear programming (MILP) model developed for the analysis of H3 histone that utilizes the total number of possible modifications, given modified parent mass, to derive the most probable set of PTMs. We will also present a generalized de novo method for the identification of modified and unmodified proteins using mixed-integer linear optimization and ETD/ECD tandem mass spectrometry.

[1] Kinter M and Sherman NE. Protein Sequencing and Identification Using Tandem Mass Spectrometry. New York: Wiley, 2000.

[2] Zubarev RA, Kelleher NL, McLafferty FW. Electron capture dissociation of multiply charged protein cations: a nonergodic process. J. Am. Chem. Soc. 120:3265-3266, 1998.

[3] Bakhtiar R and Guan Z. Electron capture dissociation mass spectrometry in characterization of peptides and proteins. Biotechnol. Lett. 28:1047-1059, 2006.

[4] Syka JEP, Coon JJ, Schroeder MJ, Shabanowitz J, Hunt DF. Peptide and protein sequence analysis by electron transfer dissociation mass spectrometry. Proc. Nat. Acad. Sci. 101(26):9528-9533, 2004.

[5] Mikesh LM, Ueberheide B, Chi A, Coon JJ, Syka JEP, Shabanowitz J, Hunt DF. The utility of ETD mass spectrometry in proteomic analysis. Biochimica. et Biophysica Acta. 1764:1811-1822, 2006.

[6] Udeshi NH, Shabanowitz J, Hunt DF, Rose KL. Analysis of proteins and peptides on a chromatographic timescale by electron-transfer dissociation MS. FEBS Journal. 274:6269-6276, 2007.

[7] Molina H, Horn DM, Tang N, Mathicanan S, Pandey A. Global proteomic profiling of phosphopeptides using electron transfer dissociation tandem mass spectrometry. Proc. Nat. Acad. Sci.

104(7):2199-2204, 2007.

[8] Garcia BA, Pesavento JJ, Mizzen CA, Kelleher NL. Pervasive combinatorial modification of histone H3 in human cells. Nat. Meth. 4(6):487-489, 2007.

[9] DiMaggio PA and Floudas CA. A mixed-integer optimization framework for de novo peptide identification. AIChE Journal. 53(1): 160-173, 2007.

[10] DiMaggio PA and Floudas CA. De novo peptide identification via tandem mass spectrometry and integer linear optimization. Anal. Chem. 79: 1433-1446, 2007.

[11] DiMaggio PA, Floudas CA, Lu B, Yates JR. A hybrid methodology for peptide identification using integer linear optimization, local database search, and QTOF or OrbiTrap tandem mass spectrometry. J. Proteome Res. 7:1584-1593, 2008.

[12] Geer LY, Markey SP, Kowalak JA, Wagner L, Xu M, Maynard DM, Yang X, Shi W, Bryant SH. Open Mass Spectrometry Search Algorithm. J. Proteome Res. 3: 958-964, 2004.

456a Multiplexed Proteomic Technologies for the Study of Host-Pathogen Interactions

Brett A. Chromy, Chemistry, Materials, Earth, and Life Sciences, Lawrence Livermore National Laboratory, 7000 East Ave., L-452, Livermore, CA 94551

New proteomic technologies that utilize multiplexing to increase sample throughput and enrich datasets as compared to individual proteomic experiments are increasingly being used for biomarker discovery and other applications. Our work has centered on using these proteomic technologies in the study of host-pathogen interactions. We have studied both the pathogen proteomes and host proteomes following response to these pathogens in human and other animal systems. This presentation will show our work with 2-D DIGE, SELDI-MS, and protein arrays in an effort to better understand these systems and discover new countermeasure targets against these human and other animal pathogens. Our work on the plague bacterium, *Yersinia pestis*, its proteome under diverse conditions and diverse host response models to this pathogen will be highlighted. The presentation will also showcase key advances in the proteomics field that have helped to capitalize on these multiplexed technologies.

A Microfluidic Device for Physical Extraction of Intracellular Proteins from Bacterial Cells

Ning Bao, Chang Lu*
Department of Agricultural and Biological Engineering, School of Chemical Engineering, Birck Nanotechnology Center, Bindley Bioscience Center, Purdue University, West Lafayette, Indiana 47907

*Corresponding author: Chang Lu, Department of Agricultural and Biological Engineering, 225 S. University Street, West Lafayette, IN 47907.
Tel: 765-494-1188
Fax: 765-496-1115
Email: changlu@purdue.edu

Abstract

The analysis of intracellular proteins from bacterial cells typically requires a cell population of ~1000 or more due to the small cell size. Effective combination of concentration and lysis steps is desired for such analysis to be carried out on a microfluidic platform. In here, we report a simple microfluidic device that integrates the capture of the bacterial cells using a microscale bead array and the rapid electrical lysis for the release of intracellular proteins. We study the retention of *Escherichia coli* cells with different concentrations in this type of bead array and the optimal electrical parameters for the electroporative release of intracellular proteins. The bead array can be replaced after each run and this allows the device to be used for multiple times. Our design provides a simple solution to the extraction of intracellular proteins from a bacterial cell population based entirely on physical methods without applying chemical or biological reagents. Our device forms a critical basis for bacterial proteomic studies based on microfluidics.

Introduction

There are numerous assays applied in biological research that are essentially based on analysis of intracellular materials from cells. Conventional assays based on nucleic acids or proteins from cells require at least hundreds to thousands of cells as the starting material. Recently new tools such as capillary electrophoresis and microfluidics have allowed researchers to analyze intracellular contents from a low number of cells [1-9]. Due to powerful amplification techniques such as polymerase chain reaction and detection techniques such as laser-induced fluorescence, the analysis of intracellular nucleic acids and proteins at the single cell level has become routine for mammalian cells [9-13]. However, the analysis of intracellular materials from bacterial cells, especially proteins, still has to be conducted based on a fairly large number of cells. Bacteria are important model organisms to study regulatory networks and protein functions because they have relatively small genomes and low number of protein components compared to eukaryotic cells [14]. Furthermore, there have been established methods for genetic manipulation of some bacteria which make the studies of protein functions attainable. Finally, bacterial proteomic studies provide crucial information on the adaptation networks that are important for bacterial survival and such knowledge is critical for the discovery of antibacterial drugs [15, 16]. For example, a single *Escherichia coli* cell has a length of 2 μm and a diameter of 0.8 μm which yield a volume of ~1 fl [17]. This volume is about three orders of magnitude smaller than that of a typical eukaryotic somatic cell. Analysis of intracellular proteins from bacterial cells typically requires the gathering of a substantial number of cells in a relative small volume to generate the desired concentration needed for detection. Effective concentration strategies can save valuable time on the enrichment (the selective expansion of the bacterial cell population) typically needed for bacteria detection and more importantly eliminate potential artifacts generated by such expansion. The capture or concentration of bacterial cells in the solution is a nontrivial task due to the small dimensions of the cells. Dielectrophoresis and electrokinetic techniques have been applied to concentrate bacterial cells on microfluidic chips [18-20]. Fabricated or membrane structures were also used for capturing bacterial cells by having feature sizes comparable to or smaller than the cell dimensions [21, 22]. On the other hand, there have been a number of lysis methods developed for the disruption of bacterial cell membrane using mechanical [23-25], chemical [9, 26], thermal [27, 28], and electrical [29-31] means. However, there have been very few reports focusing on the integration of capture and lysis of bacterial cells for analysis of intracellular contents [32].

In this work, we demonstrate a simple microfluidic device that can serve as a basic unit operation for the capture and analysis of bacterial cells. Microscale silica beads (~3.0 or 4.8 μm in the diameter) are packed in a microfluidic channel and the array of beads provides microscale matrix that filters *E. coli* cells in the solution. The packing and release of the microscale beads are achieved using a mechanical valve fabricated by multilayer soft lithography [33]. A subsequent electrical pulse rapidly lyses the cells and releases intracellular proteins from the cells. The green fluorescent protein (GFP) released by GFP expressing *E. coli* cells is detected using laser-induced fluorescence in the downstream of the channel. The bead array here is a temporary structure that can be removed by opening the valve. These features make our device a simple and versatile tool that can be easily incorporated in a microfluidic system for applications involving analysis of proteins from bacterial cells.

Experimental Section

Microchip fabrication

The multilayer microfluidic chips based on Polydimethylsiloxane (PDMS) (General Electric Silicones RTV 615, MG Chemicals, Toronto, ON, Canada) with pneumatic valves were fabricated using multilayer soft lithography with substantial modifications [33]. The microfluidic structures for the fluidic and control layers were designed using FreeHand MX (Macromedia, San Francisco, CA) and then printed out separately on transparencies with a resolution of 5080 dpi. The transparencies were used as the photomasks for the fabrication of two masters on silicon wafers. The positive photoresist AZ 9260 (Clariant, Somerville, NJ) was used for fabricating the fluidic channel master and the negative photoresist SU-8 2025 (MicroChem, Newton, MA) was used for fabricating the control channel master. The depth of the fluidic channels (the thickness of the SU-8 2025) was ~12 μm, measured by a profilometer (Sloan Dektak3 ST). The depth of the control channels was ~58 μm. The PDMS prepolymer mixture consisting of A and B with the mass ratio of 10:1 was used for the fabrication of both the fluidic and control layers. The control layer had a thickness of ~5 mm. The fluidic layer was created by spinning PDMS prepolymer mixture on the fluidic master at 1200 rpm for 35 s which yielded a thickness of 93 μm. This determined that the membrane between the control and fluidic channels had a thickness of 81 μm. The two PDMS layers were firstly cured at 80 ºC for 20 min before the channel surface of the control channel and the upper surface of the fluid channel were oxidized using a Tesla coil (Kimble/Kontes, Vineland, NJ) in atmosphere. The oxidized PDMS surfaces were immediately aligned and brought into contact and then heated in the oven at 80 ºC for 2 hr to form a two-layer PDMS chip. The access holes were punched in the PDMS chip. A glass slide was cleaned in a basic solution (H_2O: NH_4OH (27%): H_2O_2 (30%) = 5:1:1, volumetric ratio) at 85 °C for 1 hr, rinsed with DI water and then blown dry. The PDMS chip and the glass slide were oxidized using the Tesla coil in atmosphere before they were brought into contact immediately and heated at 80 °C for another 2 hr to form a closed microchip.

Bacterial culture and preparation

GFP-expressing *E. coli* cells (transformed by pQBI T7-GFP plasmid, Qbiogene, Irvine, CA) were used in the experiments. The bacterial cells were cultured in Luria-Bertani (LB) broth (BIO 101 Systems, Irvine, CA) with 50 μg/ml of ampicillin (Amresco Inc., Solon, OH) at 37 °C for 18 hr. The density of the cells after culture was ~1×10^9/ml, determined using plate count. The bacterial culture with the volume of 1 ml was centrifuged and then the supernatant was removed. The cell pellet was resuspended in 1 ml 20 mM phosphate buffer (Na_2HPO_4+NaH_2PO_4, pH=7.5). The resulting suspension was directly used or diluted by the phosphate buffer to form densities between 10^3 and 10^{11} cells/ml and then flowed into the device driven by an infusion pump (PHD, Harvard Apparatus, Holliston, MA).

Microchip operation

The device was mounted on an inverted fluorescence microscope (IX71, Olympus, Melville, NY). The layout and the setup of the microchip are shown in Figure 1. The control channel was filled with water to prevent the leakage of air into the fluidic channel. The valve was actuated by pressurizing the control channel using a fast-response solenoid valve (ASCO Scientific). A pressure regulator with a pressure meter was used to adjust the pressure in order to retain the beads by partially closing the channel. The solenoid valve was automated by a LabVIEW program. For the accumulation of silica beads with the diameter of 4.8 or 3.0 μm (Bangs lab, Fishers, IN), the pressure of around 16 psi was needed to partially close the valve with the bead suspension having a constant flow rate of 1 μl/min. We accumulated a bead array with a length of ~ 600 μm and then we switched the inlet solution to bacterial suspension with the same flow rate (1 μl/min) while the valve was kept partially closed during all time.

The detection of the released GFP from bacterial cells after electrical lysis was carried out by focusing a laser focal volume in the downstream of the bead array (~300 μm away from bead array) using a 60X dry objective (NA=0.70). A 488 nm air-cool argon ion laser (Spectra-Physics, Mountain View, CA) was used in the experiments. The excitation and emission were filtered by a 505DCLP dichroic beamsplitter (Chroma Technology Corp., Rockingham, VT). The fluorescence from the released GFP was collected by the same objective and filtered by a D535/40 emission filter (Chroma Technology Corp., Rockingham, VT) before the fluorescent signal was collected by a R9220 photomultiplier tube (PMT) (Hamamatsu Inc, Bridgewater, NJ) biased at the voltage of –1100 V. The photocurrent from the PMT was amplified and filtered by a Low Noise Current Preamplifier (SR570, Standard Research System, Sunnyvale, California). The amplified current was converted to voltage and then digitalized by a PCI data acquisition card (PCI-6254, National Instruments, Austin, TX). The electrical pulses (3 s pulses with 10 s intervals between the pulses) for cell lysis was generated by a high voltage power supply (1AA12P30, Ultravolt, Ronkonkoma, NY) controlled by a LabView program. The fluorescent signal from the released GFP was collected while the pulses were applied.

The fluorescent and optical images were taken using ORCA-285 CCD camera (Hamamatsu, Bridgewater, NJ). A 100-W mercury lamp was applied to generate the excitation and the fluorescence was filtered by a filter cube consisting of exciter HQ480/40, emitter

HQ535/50 and beam splitter Q505lp (Chroma Technology, Rockingham, VT).

Results and discussion

In this work, we created an array of microscale beads in a microfluidic channel by pinching the PDMS channel using a mechanical valve made by multilayer soft lithography with modifications [9, 33]. When we made the master for the fluidic layer, we intentionally avoided the reflow of the photoresist under high temperature. Such reflow was critical for producing the round cross-sectional profile of the fluidic channel so that the complete closure of the channel by such a valve is possible [33]. In our application, we had only the center part of the PDMS membrane in contact with the glass bottom when we actuated the valve due to the square cross section of the fluidic channel (as shown in the inset image of Figure 1). The partial closure of the valve allowed the flow of the fluid and the retention of the beads simultaneously. The advantage of creating the array using the mechanical valve is that the bead array can be formed and removed at will by partially closing or completely opening the valve. Such design allows the device to have multiple purposes other than the capture and analysis of bacterial cells. When the valve was pinched under a pressure of 16 psi, the beads with 4.8 μm diameter were retained by the partially closed valve with a flow rate of 1 μl/min in the channel and accumulated to form a column. Due to the depth and the cross section of the channel (~200 μm × 12 μm), 2~3 layers of the beads were packed in the channel. Assuming the beads are closely packed, the gaps between the beads (potentially allowing the passage of spheres with diameters of 0.75 to 1.25 μm, depending on the fashion that the beads are packed) would allow possible passage of *E. coli* cells (averagely 0.8 μm in the diameter of the rod shape) with difficulty created by the trapping in the gaps.

We tested the capture of *E. coli* cells from a solution with a density of $\sim 1\times10^3$ cells/ml (diluted by 10^6 times from the original culture broth) at a flow rate of 1 μl/min through the channel. We observed that the cells were all captured by the bead array. The cells penetrated deeper into the array as time progressed and the cells were distributed through the array with a fairly uniform density (Figure 2a). With the low concentration of the cells in the bead array, the number of the fluorescent spots roughly corresponded to the number of cells. The plot of the number of the cells over time (Figure 2b) indicates that the capture of the cells was close to 100%.

We further increased the density of *E. coli* cells to 1×10^7 cells/ml. The images in Figure 3a show that the bead array became heavily populated with *E. coli* cells over the time. We were able to estimate the cell concentration in the array based on the calibration of the fluorescence intensity against fluorescent images from bacterial suspension with known concentrations (Supporting Information Figure S1). The plot of the number of cells against the time in Figure 3b shows that the array was able to concentrate the cells by a factor of 10^4 within 40 min. Based on our data, the 4.8 μm bead array with the cross section of 200 μm×12 μm can accommodate up to $\sim 1\times10^5$ cells/100 μm length.

The retention of the *E. coli* cells is due to the physical trapping in the microscale structures in the bead matrix when the gaps in the 4.8 μm-diameter-bead matrix (>0.75 μm) were slightly larger than the cell size. The dynamic movement of *E. coli* cells in the matrix was observed while the overall occupancy by the cells gradually increased over time (Supporting Information Movie S1). This mechanism is purely physical and very different from commonly used immunoseparation based on antigen-antibody interaction [34, 35]. Since the identity of the bacterium in proteomic studies is often known, our approach offers a simple alternative for bacteria capturing and also avoids interference with the results due to introducing an antibody.

Since the size of the gaps between the beads affects the retention of the cells, we also applied silica beads with the diameter of 3.0 μm. The gaps between the beads in this case will allow the entry of particles with a cross-sectional diameter from 0.46 to 0.71 μm. As expected, due to the smaller gaps between the beads, we observed that *E. coli* cells were stopped at the edge without penetrating into the bead array (Supporting Information Figure S2). The bacterial cell layer accumulated over time and such cell layer apparently did not block the flow of the solution. The cells in solution were captured with an efficiency of 100%. The result indicates that 3.0 μm bead array may be useful for applications requiring very high efficiency collection of bacterial cells from the solution. On the other hand, the disadvantage of 3.0 μm bead array compared to 4.8 μm bead array is that it requires a higher pressure to drive the flow due to the array's higher volumetric occupancy and this demands higher performance of the multilayer valve.

The microfluidic bead array provides an ideal platform for the incorporation of rapid lysis methods such as electrical lysis after cell capture. Electrical lysis is generally considered to be the most rapid lysis method and does not require chemical or biological reagents which may interfere with subsequent assays. Electrical lysis of bacterial cells has been applied to flowing cells by us and other researchers [30, 31]. However, in these settings it was difficult to generate cell lysate with substantial concentration for proteomic analysis since the flowing cells could not be easily concentrated. In this study, we accumulated GFP-expressing *E. coli* cells with an original concentration of ~10^7 cells/ml concentrated in the 4.8 μm bead array for around 40 min. We focused a laser in the downstream of the bead array and applied periodic electrical pulses (3 s duration for each pulse with 10 s intervals between pulses) in the channel with different field intensities after such cell capture. We were able to detect fluorescence bursts after the pulses when the field intensity was higher than 1000 V/cm (Figure 4). This threshold is similar to what we observed for electrical lysis of flowing cells [31]. The fluorescence burst indicates the release of intracellular materials (GFP and other intracellular materials generating autofluorescence). Such release was mostly finished within the first electrical pulse when the field intensity was at 1250 V/cm and occurred slower (within the first 6-7 pulses) when the field intensity was at 1000 V/cm. These results reveal that when combined with electrical lysis, the device can produce cell lysate from bacterial cells after the capture of cells using the bead array. The release of intracellular GFP was also witnessed by fluorescent images before and after the lysis and the movie taken during the process of the pulse application (Supporting Information Figure S3 and Movie S2).

One important feature of the device is its reusability after cell lysis. The bead array together with lysed cells can be flushed out when the mechanical valve is open after one run. The bead array can be formed again and the device can be reused after that. We observed that too many electrical pulses at high intensity (e.g. >20 pulses with each lasting 3s at 1000 V/cm) could create dents in the PDMS membrane (Supporting Information Figure S4). The damage at 1250 V/cm was visually similar. This effect of the electrical field on PDMS has been reported previously [36]. The dents decreased the mechanical strength of the valve and led to higher pressure required for the actuation of the valve. This problem can be solved by having multiple valves in the channel and using the valves one by one, when more than 20-30 runs are desired on one device.

Conclusions

In this study, we demonstrate using microscale bead arrays to capture *E. coli* cells. The bead array (formed by silica beads of 3.0 or 4.8 μm diameter) is created using a pneumatic valve which also allows the removal of the structure after operation. The bead array provides the matrix needed for capturing *E. coli* cells in the solution. Electrical lysis is also incorporated to produce rapid release of intracellular proteins from captured cells. Both the bacteria capturing and the lysis are carried out based on physical methods. The approach is simple and eliminates the application of chemical and biological reagents which may alter the results in proteomic studies.

Acknowledgement

The authors acknowledge support by a cooperative agreement with the Agricultural Research Service of the US Department of Agriculture, project number 1935-42000-035, through the Center for Food Safety Engineering at Purdue University.

Supporting Information Available

Additional information as noted in text. This material is available free of charge via the Internet at http://pubs.acs.org.

References

1. Kennedy, R. T.; Oates, M. D.; Cooper, B. R.; Nickerson, B.; Jorgenson, J. W. *Science* **1989,** 246, 57-63.
2. Olefirowicz, T. M.; Ewing, A. G. *Anal. Chem.* **1990,** 62, 1872-1876.
3. Hogan, B. L.; Yeung, E. S. *Anal. Chem.* **1992,** 64, 2841-2845.
4. Krylov, S. N.; Zhang, Z. R.; Chan, N. W. C.; Arriaga, E.; Palcic, M. M.; Dovichi, N. J. *Cytometry* **1999,** 37, 14-20.
5. Meredith, G. D.; Sims, C. E.; Soughayer, J. S.; Allbritton, N. L. *Nat. Biotechnol.* **2000,** 18, 309-312.
6. Dovichi, N. J.; Hu, S. *Curr. Opin. Chem. Biol.* **2003,** 7, 603-608.
7. McClain, M. A.; Culbertson, C. T.; Jacobson, S. C.; Allbritton, N. L.; Sims, C. E.; Ramsey, J. M. *Anal. Chem.* **2003,** 75, 5646-5655.
8. Wang, H. Y.; Lu, C. *Chem. Commun.* **2006**, 3528-30.
9. Hong, J. W.; Studer, V.; Hang, G.; Anderson, W. F.; Quake, S. R. *Nat. Biotechnol.* **2004,** 22, 435-439.
10. Kennedy, R. T.; Oates, M. D.; Cooper, B. R.; Nickerson, B.; Jorgenson, J. W. *Science* **1989,** 246, 57-63.
11. Zhang, Z.; Krylov, S.; Arriaga, E. A.; Polakowski, R.; Dovichi, N. J. *Anal. Chem.* **2000,** 72, 318-322.
12. Hu, S.; Michels, D. A.; Fazal, M. A.; Ratisoontorn, C.; Cunningham, M. L.; Dovichi, N. J. *Anal. Chem.* **2004,** 76, 4044-4049.
13. Li, H.; Yeung, E. S. *Electrophoresis* **2002,** 23, 3372-80.
14. Noirot, P.; Noirot-Gros, M. F. *Curr. Opin. Microbiol.* **2004,** 7, 505-12.
15. Brotz-Oesterhelt, H.; Bandow, J. E.; Labischinski, H. *Mass Spectrom. Rev.* **2005,** 24, 549-565.
16. Nilsson, C. L. *Am. J. Pharmacogenomics* **2002,** 2, 59-65.
17. Stanier, R. Y.; Ingraham, J. L.; Wheelis, M. L.; Painter, P. R., *The microbial world.* 5 ed.; Prentice-Hall: Englewood Cliffs, NJ, 1986.

18. Cabrera, C.; Yager, P. *Electrophoresis* **2001,** 22, 355-362.
19. Cheng, I.-F.; Chang, H. C.; Hou, D.; Chang, H.-C. *Biomicrofluidics* **2007,** 1, 021503.
20. Zhou, R. H.; Wang, P.; Chang, H. C. *Electrophoresis* **2006,** 27, 1376-1385.
21. Floriano, P. N.; Christodoulides, N.; Romanovicz, D.; Bernard, B.; Simmons, G. W.; Cavell, M.; McDevitt, J. T. *Biosens. Bioelectron.* **2005,** 20, 2079-2088.
22. He, B.; Tan, L.; Regnier, F. *Anal. Chem.* **1999,** 71, 1464-1468.
23. Belgrader, P.; Hansford, D.; Kovacs, G. T. A.; Vankateswaran, K.; Mariella, R.; Milanovich, F.; Nasarabadi, S.; Okuzumi, M.; Pourahmadi, F.; Northrup, M. A. *Anal. Chem.* **1999,** 71, 4232-4236.
24. Kim, J.; Jang, S. H.; Jia, G. Y.; Zoval, J. V.; Da Silva, N. A.; Madou, M. J. *Lab Chip* **2004,** 4, 516-522.
25. Taylor, M. T.; Belgrader, P.; Furman, B. J.; Pourahmadi, F.; Kovacs, G. T. A.; Northrup, M. A. *Anal. Chem.* **2001,** 73, 492-496.
26. Schilling, E. A.; Kamholz, A. E.; Yager, P. *Anal. Chem.* **2002,** 74, 1798-1804.
27. Liu, R. H.; Yang, J. N.; Lenigk, R.; Bonanno, J.; Grodzinski, P. *Anal. Chem.* **2004,** 76, 1824-1831.
28. Waters, L. C.; Jacobson, S. C.; Kroutchinina, N.; Khandurina, J.; Foote, R. S.; Ramsey, J. M. *Anal. Chem.* **1998,** 70, 158-162.
29. Cheng, J.; Sheldon, E. L.; Wu, L.; Uribe, A.; Gerrue, L. O.; Carrino, J.; Heller, M. J.; O'Connell, J. P. *Nat. Biotechnol.* **1998,** 16, 541-546.
30. Lee, S. W.; Tai, Y. C. *Sens. Actuators A* **1999,** 73, 74-79.
31. Wang, H. Y.; Bhunia, A. K.; Lu, C. *Biosens. Bioelectron.* **2006,** 22, 582-588.
32. Lagally, E. T.; Lee, S. H.; Soh, H. T. *Lab Chip* **2005,** 5, 1053-8.
33. Unger, M. A.; Chou, H. P.; Thorsen, T.; Scherer, A.; Quake, S. R. *Science* **2000,** 288, 113-116.
34. Kaclikova, E.; Kuchta, T.; Kay, H.; Gray, D. *J. Microbiol. Methods* **2001,** 46, 63-67.
35. Ho, J. A.; Hsu, H. W.; Huang, M. R. *Anal. Biochem.* **2004,** 330, 342-9.
36. McDonald, J. C.; Metallo, S. J.; Whitesides, G. M. *Anal. Chem.* **2001,** 73, 5645-5650.

Figure Captions

Figure 1. A schematic of the microfluidic device for intracellular analysis of bacterial cells. The most important parts of the device included the microscale bead array and a two-layer pneumatic valve that formed or removed the bead array. The width and length of the fluidic channel were 200 μm and 10 mm, respectively. The width of the control channel was 200 μm. The inset image shows the formed bead array by partially closing the valve.

Figure 2. The capture of *E. coli* cells at the concentration of $\sim 1\times10^3$ cells/ml by the bead array. The bead array was formed by silica beads with the diameter of 4.8 μm. (a) The optical image of the bead array and the fluorescent images of captured cells over time (at 30, 60, and 90 min). (b) The number of cells captured over time.

Figure 3. The capture of *E. coli* cells at the concentration of $\sim 1\times10^7$ cells/ml by the bead array. The bead array was formed by silica beads with the diameter of 4.8 μm. (a) The optical image of the bead array and the fluorescent images of captured cells over time (at 2, 10 and 60 min). (b) The variation in the cell concentration in the bead array observed over time. The cell concentration at different times (indicated by the horizontal lines) was estimated by comparing the fluorescence intensity of the images to those of the images with known concentrations (Figure S1 in Supporting Information).

Figure 4. The fluorescence intensity detected in the downstream of the bead array under different lysis field intensities when the series of electrical pulses were applied. Each trace was generated by a fresh bead array (formed by 4.8 μm beads) and captured cells (~1×10^7 cells/ml solution flowing for around 40 min). The pulse sequence consisted of 15 pulses of 3s duration with 10s intervals in between for pulse intensities at 500, 750, and 1000 V/cm. The pulse sequence consisted of 5 pulses of the same pattern at 1250 V/cm.

Figures

Figure 1

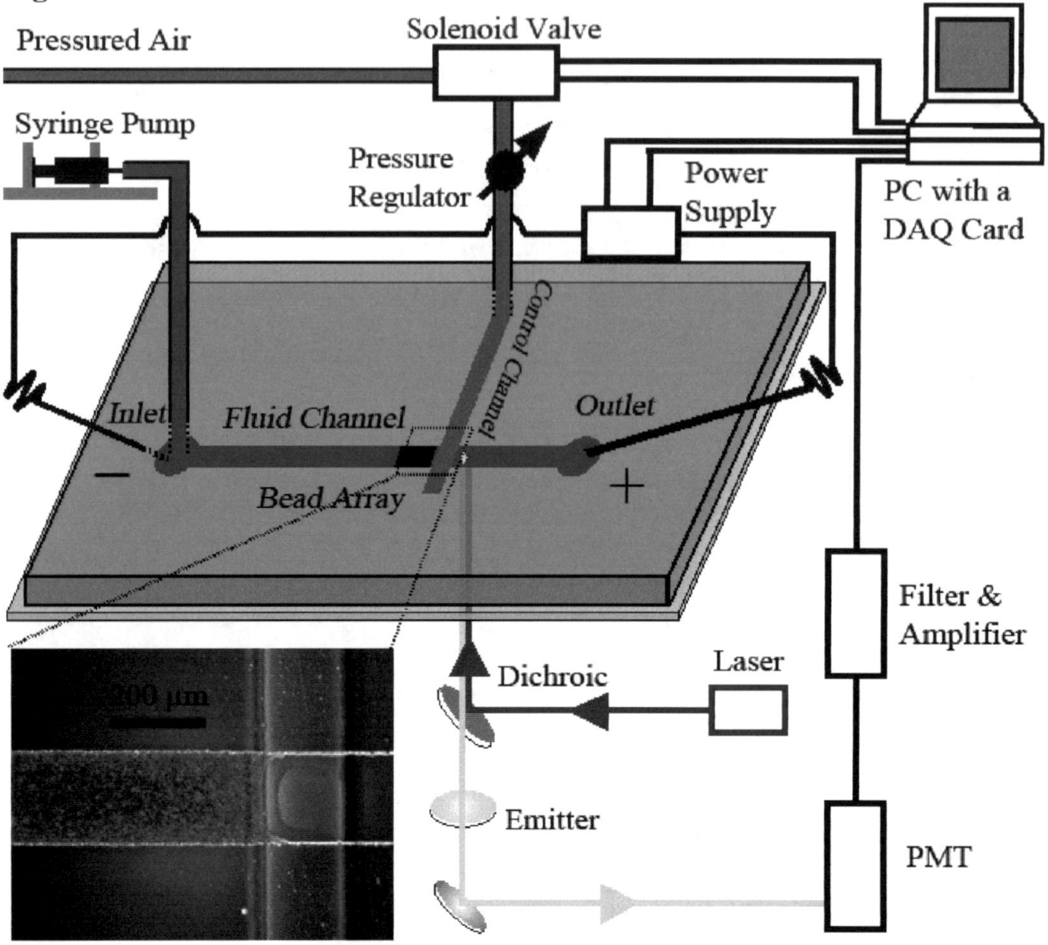

Figure 2

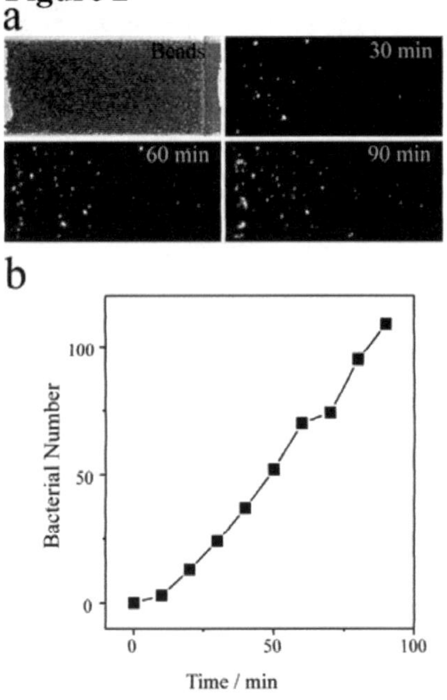

Figure 3

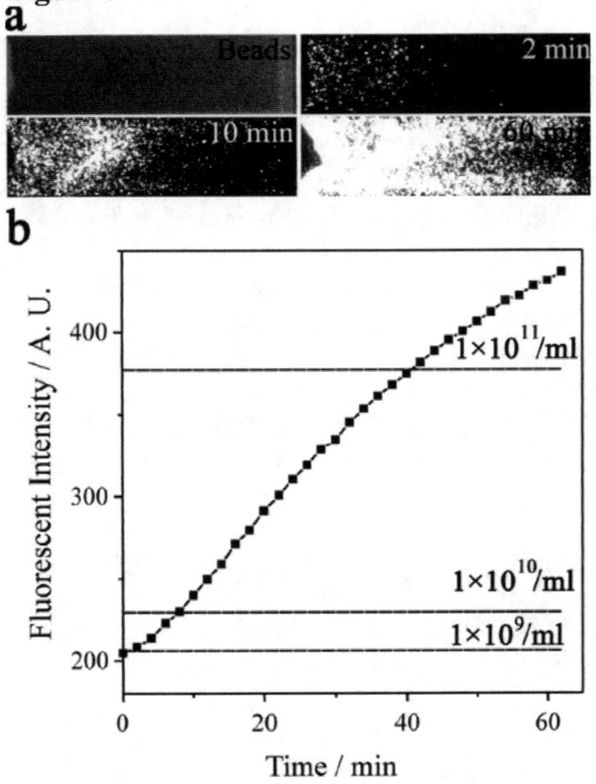

Figure 4

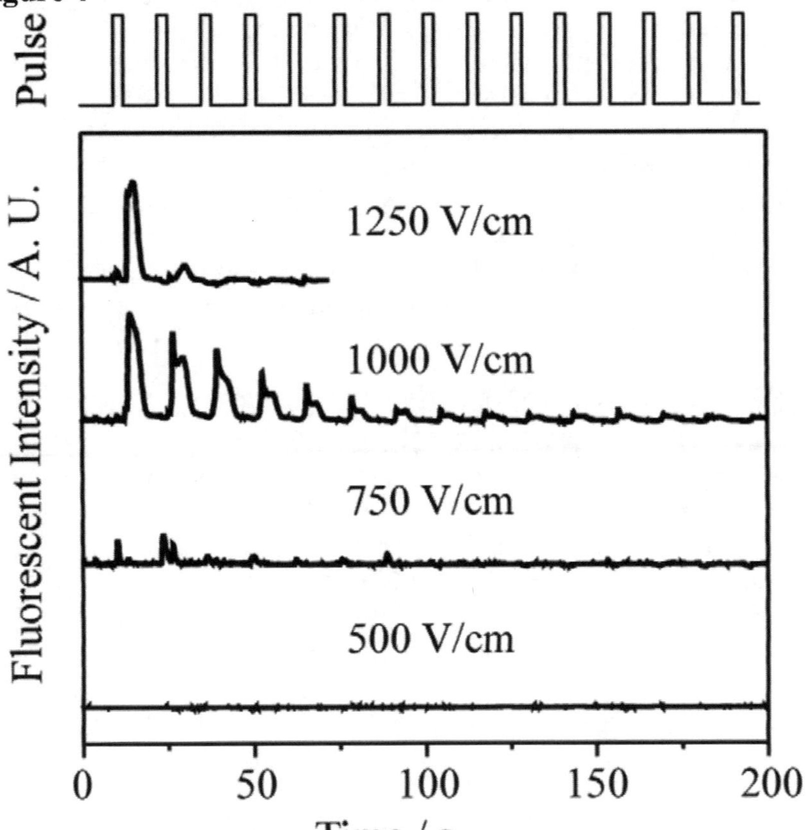

Eliminating Electrokinetic Cross-Talk in Nano-Channel Arrays for Biomolecular Detection

We have designed and tested nanochannel sensors for biomolecular detection using fundamental analysis of the underlying electrokinetic phenomena. These nanochannels are bounded by two micro-reservoir to allow high throughput. A different molecular probe (oligos for DNAs, antibodies for proteins) is functionalized onto the surface of each nanochannel to capture a specific molecular target that is driven electrophoretically, electro-osmotically or dielectrophoretically into the channel. Due to the small dimension of the channel, molecular capture can significantly change the complex cross-channel impedance signal, detected by the same microelectrodes that drove the molecules into the nanochannels. We find, however, that the impedance of a single channel is also sensitively dependent on the depletion and enrichment zones at the two ends of the nanoelectrodes. Both zones are produced by the conductivity jump across the nano/micro junction as double layer overlap in the nanochannels produce a higher ionic strength. We extend Dukhin, Rubinstein and our earlier one-dimensional steady-state theory for the extended polarized and diffusion layers of the depletion zone to AC conditions relevant to impedance measurements (Dukhin 1991, Ben & Chang 2002, Rubinstein & Zaltzman 2000). Furthermore, our experiments extend previous studies on DC concentration polarization effects at micro-nano junctions (e.g. Pu et al. 2004, Plecis et al. 2005, Kim et al. 2007) to AC conditions. Quite curiously, we find that the formation dynamics of these layers are diffusive in nature and are hence universal - frequency dependent. However, by designing channels with different dimensions, we can produce an internal impedance that is higher than the generic entrance impedance, thus endowing each channel with a different impedance signature with and without captured molecules. As a result, a multi-channel array can be designed to allow detection of a large number of targets, provided electric cross-talk does not occur between the channels. By analyzing the dynamics and extent of the depletion/enrichment layer for each pore, we eliminate this possibility by selecting the right frequency and channel spacing such that neighboring depletion layers do not overlap. This involves depletion layer formation in different geometries, with unique impedance signals of their own that can be monitored to ensure lack of cross-talk.

References:

[1] Y. Ben and H.-C. Chang, Nonlinear Smoluchowski slip velocity and micro-vortex generation, *J. Fluid Mech.*, 461, 229-238 (2002).

[2] S.S. Dukhin, Electrokinetic phenomena of the second kind and their applications, *Advances in Colloid and Interface Science*, 35, 173-196 (1991).

[3] I. Rubinstein and B. Zaltzman, Electro-osmotically induced convection at a permselective membrane, *Physical Review E*, 62, 2 (2000).

[4] S.J. Kim, Y.C. Wang, J.H. Lee, H. Jang, and J. Han, Concentration Polarization and Nonlinear Electrokinetic Flow near a Nanofluidic Channel, *PRL* 99, 044501 (2007).

[5] A. Plecis, R.B. Schoch, and P. Renaud, Ionic Transport Phenomena in Nanofluidics: Experimental and Theoretical Study of the Exclusion-Enrichment Effect on a Chip, *Nano Lett.*, 5, 1147-1155 (2005).

[6] Q. Pu, J. Yun, H. Temkin, and S. Liu, Ion-Enrichment and Ion-Depletion Effect of Nanochannel Structures, *Nano Lett.*, 4, 1099-1103 (2004).

456d The Use of a Bead-Based Combinatorial Peptide Library to Enrich Low-Abundance Proteins from Various Sample Types

Tom Berkelman and Steve Freeby. Life Science Group, Bio-Rad Laboratories, 6000 James Watson Dr., Hercules, CA 94547

One of the fundamental challenges in proteome analysis is the fact that samples are often dominated by a relatively small number of high-abundance proteins whose presence can obscure less abundant proteins and limit the capacity and resolution of the separation technique employed. Samples are therefore often treated to remove high-abundance proteins. ProteoMiner™ protein enrichment is a novel technology which uses a bead-based library of combinatorial peptide ligands, which act as unique binders for proteins. When a complex sample is applied to the beads, high-abundance proteins will saturate their specific ligands and excess protein will be washed away. Low-abundance proteins will be concentrated on their specific ligands and following elution, the low-abundance proteins will be enriched relative to the high-abundance proteins.

ProteoMiner protein enrichment has been most extensively applied to serum and plasma, and the commercially available kit has been optimized for these samples. However, one of the main potential strengths of the ProteoMiner technique is its lack of intrinsic specificity for any particular sample type. ProteoMiner can in theory be used to decrease high-abundance proteins in any sample that could benefit from such a treatment. Cell- or tissue-derived samples can differ widely in protein concentration, protein abundance distribution and the presence of interfering substances. The basic method therefore needs to be tailored somewhat for each sample type. Progress in developing methods for applying ProteoMiner to diverse sample types will be presented.

456e 2D Electrophoresis: Isotachophoresis Followed by Isoelectric Focusing

Bingwen Liu and Cornelius F. Ivory. School of Chemical Engineering and Bioengineering, Washington State University, Pullman, WA 99163

The primary challenge to proteomic analysis of complex biological fluids is the vast diversity of proteins, >10^5, and their broad dynamic range, >10^8. These issues can be simultaneously addressed by developing separation platforms with high peak capacities and which can concentrate low-abundance proteins (1 mg/ml). Since no single separation technique can accomplish this task alone, it is necessary that several orthogonal "dimensions" of separation be multiplexed into a single platform.

This paper describes a 2D nonlinear electrophoresis system on a PDMS chip which combines isotachophoresis (ITP) followed by isoelectric focusing (IEF). ITP is used as the first dimension to preconcentrate low-abundance proteins in the sample. The concentrated sample zones are then fed into channel segments through which IEF is applied to separate them based on differences in their isoelectric points (ΔpI). A mixture of three fluorescent proteins is used to demonstrate that they are successfully resolved into three bands in the IEF dimension after ITP preconcentration. Further integration of a third dimension is then possible after IEF.

456f a Generic Platform for Label-Free Detection of Proteins, DNA, and Other Charged Analytes

Yu-Wen Huang, Faisal Shaikh, and Victor M. Ugaz. Artie McFerrin Department of Chemical Engineering, Texas A&M University, 3122 TAMU, College Station, TX 77843

We investigate a technique to enable label-free detection of charged biomolecules that harnesses a simple microfluidic device design in which a closely spaced array of individually addressable electrodes is patterned along the floor of a microchannel. When a solution of biomolecules is loaded into the microchannel, the electrode array can be activated in a way that promotes electrophoretic transport and accumulation of biomolecules into localized zones of very high concentration. With proper design, it is possible to achieve a sufficient degree of compaction to induce formation of a concentrated mesophase that becomes clearly visible under white light. This phenomenon is possible because the closely spaced electrodes within the array (~ 200 um) enable electric fields to be generated that are strong enough (50–100 V/cm) to confine the biomolecules at the electrode surface by applying very low potentials (~1–2 V). In this way, the biomolecules experience multi-dimensional focusing and compaction into an ultra-concentrated layer that is only a few microns thick. Here we extend our previous DNA-based studies to proteins using egg white lysozyme as a model system. Using a 4% w/v initial concentration, we were able to collect the sample at an electrode with sufficient enrichment to make it visible under white light illumination with no modification to our current device design. The sample becomes visible in 10 – 20 s after applying the voltage, and dissipates 5 – 10 s after the voltage was switched off. We characterize this process and further show that this approach can be used as a generic label-free method to concentrate and detect small molecule analytes in portable bioanalysis systems without the use of chemical fluorophores.

456g Cys Shotgun Labeling to Probe Protein Unfolding and Assembly In Solution and In Cells

Christine Carag[1], David Pajerowski[1], Karthik Rajagopal[1], and **Dennis E. Discher**[2]. (1) University of Pennsylvania, Room 129 Towne Building, 220 South 33rd Street, Philadelphia, PA 19104-6393, (2) Chemical and Biomolecular Engineering, University of Pennsylvania, Room 129 Towne Building, 220 South 33rd Street, Philadelphia, PA 19104-6393

Cysteine is a relatively hydrophobic amino acid and is often buried within protein folds or assemblies, but it is rarely used to probe protein structure. For intracellular proteins, Cys are not tied up in disulfides, and so their reactivity to many fluorescent dyes in solution is readily determined as a function of temperature under both native and high urea conditions. When a protein is folded at low temperature, the reaction rate tends to be faster in urea, and the ratio of kinetics yields a thermal unfolding curve that appears very similar to results from other solution methods, such as CD. Unlike these other methods, however, the Cys labeling approach can also be applied to intact cells with membrane-permeable, cell-viable dyes at low dose. Labeling studies of stem cells, lung cancer cells, muscle cells, and additional cells treated under two different conditions, such as adherent or trypsin-released, are followed by electrophoretic separation and detailed LC/MS-MS study. What results is an identification of labeled proteins as well as labeled sites in many proteins, including disease related structural proteins such as Lamin-A/C and Dystrophin. The Cys shotgun method thus provides unique, proteomic scale measurements of protein structural changes in both solution and intact cells.

(Science 317, 663-666, 2007)

ELECTROKINETIC VELOCITY CHARACTERIZATION OF MICROPARTICLES IN GLASS MICROCHANNELS

José I. Martínez-López, Blanca H. Lapizco-Encinas, Héctor Moncada-Hernández, Marco Rito-Palomares and Sergio O. Martínez-Chapa

Tecnológico de Monterrey, Monterrey, México

Abstract

Insulator-based dielectrophoresis (iDEP) is an efficient technique with great potential for miniaturization. It has been applied successfully for the manipulation and concentration of a wide array of particles, including bioparticles such as macromolecules and microorganisms. When iDEP is applied employing DC electric fields, other electrokinetic transport mechanisms are present: electrophoresis and electroosmotic flow. In order to achieve dielectrophoretic trapping of bioparticles, dielectrophoresis has to overcome electrokinetics (electroosmosis and electrophoresis). Therefore, to improve and optimize iDEP-based separations, it is necessary to characterize these electrokinetic mechanisms under the operating conditions employed for dielectrophoretic separations. The main objective of this work was to identify the operating conditions that will benefit dielectrophoretic trapping and concentration of particles when electrokinetics is present.

This study presents the estimation of the electrokinetic mobility of microparticles suspended inside a microchannel. Micro Particle Image Velocimetry (μ*PIV*) was employed to measure the velocity of 1-μm-diameter inert polystyrene particles suspended in a 3-cm-long, 10-μm-deep, 1-mm-wide, straight glass microchannel. A parametric study was carried out by varying the properties of the suspending medium (conductivity and pH) as well as the magnitude of the applied DC electric field. The results obtained using μPIV allowed to identify the conditions under which the electrokinetic force (i.e. particle velocity) is lowest, i.e., optimal conditions for dielectrophoretic trapping. It was shown that high conductivity and low pH values for the suspending medium produce lower electrokinetic mobilities, *i.e.*, low electrokinetic force, thus benefiting dielectrophoretic trapping. These findings were proved by carrying out dielectrophoretic trapping of microparticles employing a glass microchannel that contained cylindrical insulating structures. The results obtained in this study will provide with guidelines for the optimization of iDEP-based separations.

Introduction

Dielectrophoresis (DEP) is an electrokinetic transport mechanism in which a force is exerted on a particle when it is subjected to a non-uniform electric field. DEP has been used successfully for the manipulation and concentration of a wide array for bioparticles, from macromolecules to parasites (Ozuna-Chacon *et al.*, 2008). Majority of these studies have used electrode-based DEP, where nonuniform electric field is generated employing an array of electrodes and AC fields (Markx *et al.*, 1994; Washizu, 1995; Rousselet *et al.*,1998). However, there are some drawbacks with this approach: high cost of electrode construction, complex fabrication processes and decrease of functionality due to fouling effects. The use of insulators rather than electrode arrays to produce non-uniform electric field has some advantages: they retain their functionality despite surface fouling; in addition insulators can be made from a variety of materials, including plastics, which facilitates the fabrication of equipment to handle higher flow rates. Additionally, a DC field can be used for solution and particle flow through the device by electroosmotic flow, which eliminates the need of a micro-pump (Lapizco-Encinas *et al.*, 2004a). There is growing number of studies that employ insulator-based DEP (iDEP). In iDEP the nonuniform electric field is obtained by employing arrays of insulating structures and only two electrodes. The electric field is applied along an array of micro-insulating structures creating zones of higher and lower electric field intensity throughout the array, creating dielectrophoretic traps of several biological particles like DNA (Chou *et al.* 2002), protein (Lapizco-Encinas *et al.*, 2008), yeast cells (Zhou *et al.*, 2002; Suehiro *et al.* ,2003) and microorganisms (Lapizco- Encinas *et al.*,2004a; Lapizco-Encinas *et al.*, 2004b;). When iDEP is applied through DC voltage other transport mechanisms must be considered: electrophoresis and electroosmotic flow. Electrokinesis is the superposition of electrophoresis and electroosmosis. In order to achieve dielectrophoretic trapping and concentration of particles, DEP must overcome electrokinesis and pressure-driven flow, if present. Suspending medium properties have a strong influence on the magnitude of electrokinetic force (Kirby and Hasselbrink, 2004).

In this work an examination of the optimal conditions of the suspending medium (pH and conductivity) for the dielectrophoretic trapping is made through by measuring the electrokinetic mobility of microparticles suspended inside a microchannel. Microparticle image velocimetry (*μPIV*) was used to determinate the velocity of 1-μm-diameter polystyrene carboxylated microspheres suspended inside in a 3-cm-long, 1-mm-wide and 10-μm-deep microchannel made from glass. From this measurement was possible to determine the operating conditions at which electrokinetic force is the lowest, thus, benefiting dielectrophoretic trapping. These results were confirmed by carrying out dielectrophoretic trapping of the same microparticles employing a glass microchannel containing an array of cylindrical insulating posts.

Theory

Electrokinetic phenomena are related to the formation of an electrical double layer (EDL) as result of the interaction of an aqueous solution with the static charges in dielectric surfaces. There is a conventionally introduced slipping plane that separates mobile fluid from fluid that remains attached to the surface. Electric potential at this plane is called electrokinetic potential or ζ zeta potential. Usually this potential is used to estimate the degree of charge of the EDL and it is related to the thickness of the Debye Length, Equation 1 shows the relation between the zeta potential and the electroosmotic mobility (Kuhn and Hoffstetter-Kuhn, 1993).

$$\varsigma = -\frac{\mu_{eo}\eta}{\varepsilon_0 \varepsilon_m} \qquad (1)$$

where η is the medium viscosity, ε_0 is permittivity of free space and ε_m is the relative permittivity of the suspending medium. The electrokinetic velocity is directly related to the electric field E and stands for the effects of the electroosmotic mobility μ_{eo} and electrophoretic mobility μ_{ep} as shown below:

$$V_{ek} = \mu_{ek}E \qquad (2)$$

$$\mu_{ek} = \mu_{eo} + \mu_{ep} \tag{3}$$

When relative large particles are employed (diameter > 1 μm) with a low surface charge, it is possible to neglect the effect of the electrophoretic mobility (Equation 4), thus we can consider electroosmotic mobility and electrokinetic mobility as equal.

$$\mu_{ek} \approx \mu_{eo} \tag{4}$$

Electroosmotic flow can be defined as the motion of liquid over a charged solid surface as a response of an applied voltage (Kirby and Hasselbrink, 2004). The source of the charge comes from the solid-electrolyte interaction. In glass, a negatively charged surface is produced by the deprotonation of the silanol groups forming an EDL. Equation (5) shows the linearity (first order equation) of the electroosmotic flow with the applied electric field *E*.

$$V_{eo} = \mu_{eo} E \tag{5}$$

Particle image velocimetry (*PIV*) is a well known technique employed for the measuring of particle's velocity by capturing images (Adrian, 1984). This theory applied to traditional photography was modified by Westerweel (Westerweel, 1997) to include digital images. *PIV* has evolved from macroscopic world to the microscale preserving its principles of operation. Images of particles in motion are obtained with known space delays that allow the calculation of velocity vectors from measured distances (Raffel *et al.*, 1998). However, illumination and other noise sources could interfere with particle tracing. With this limitation, *PIV* allows quantifying the average velocity of a group of particles in regions identified as interrogation windows, which generates displacement vectors per interrogation window (Prasad, 2000).

Materials and Methods

A set of experiments was carried out to measure the electrokinetic velocity of 1-μm-diameter carboxylated polystyrene microspheres by varying the suspending medium properties and the magnitude of the applied DC electric field. From this data, the particles' electrokinetic mobility and the zeta potential of glass microchannel were determined. Figure 1 show the procedure followed to obtain the electrokinetic mobility from the electrokinetic velocity using *μPIV*.

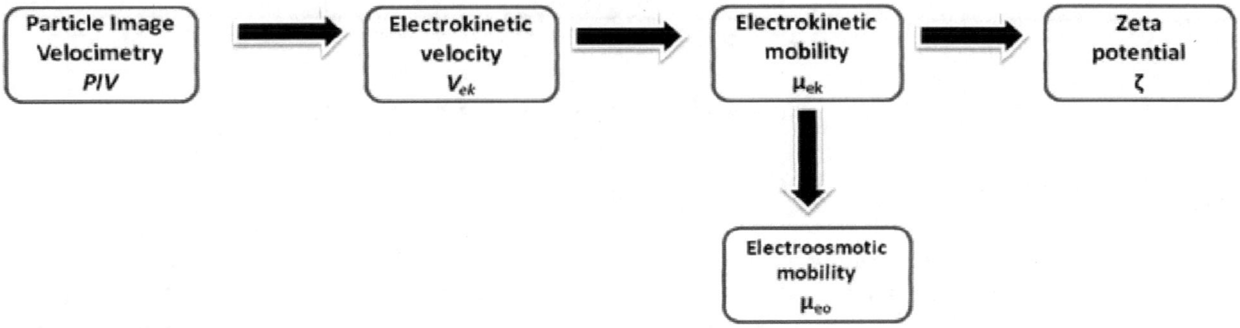

Figure 1: Methodology for obtaining electrokinetic mobility from *μPIV* measurements.

Figure 2 is a schematic representation the microdevice utilized for the electrokinetic mobility determination. This device was a glass microchannel 30 mm x 1 mm x 10 µm (l x w x d), made from two Schott 263D glass wafers (Howard Glass, Worcester, MA, USA). The particles employed were fluorescent carboxylated green-yellow 1-µm-diameter microspheres (Invitrogen, Carlsbad, CA, USA), with a concentration of 7.639 x 10^{10} microspheres/ml. These particles contain a fluorescent dye with excitation/emission of 505/515 nm.

Twelve suspending mediums were prepared adjusting pH to 6, 7, 8 and 9, and conductivity of 25, 50, 100 µS/cm, they were prepared from bidestilated water by adding NaOH and KH_2PO to adjust pH and conductivity to desired value using a pH and conductivity meter, HI255 from Hanna Instruments (Ann Arbor, MI,USA). The image and video capture device utilized for *PIV* was an inverted microfluidic microscope SVM 340 by LabSmith (Livermore, CA, USA). The microscope's control and *PIV* measurements were made with µScope software, which was provided with the microscope. A High power supply HVS448 (Livermore, CA, USA) was also used to apply DC electric fields.

Before each experiment the microchannel was cleaned with bidistillated water and then one of the 12 solutions was introduced into the channel. Pipette tips were placed at the channel inlet, to work as reservoirs, 80 µl of sample was introduced at the inlet reservoir and flow was stabilized to eliminate pressure-driven flow. Four different electric fields were applied with each suspending medium: 50, 100, 200 and 300 V/cm, each experiment was repeated twice, *i.e.* 12 videos of 30 fps were recorded for each one of the 12 suspending mediums employed. The *PIV* analysis was performed at the central region of the microchannel with a 256 x 256 pixels interrogation region using a cross correlation algorithm. The analysis of videos with *PIV* produced the values of particles velocities as a function of the suspending medium and applied electric field.

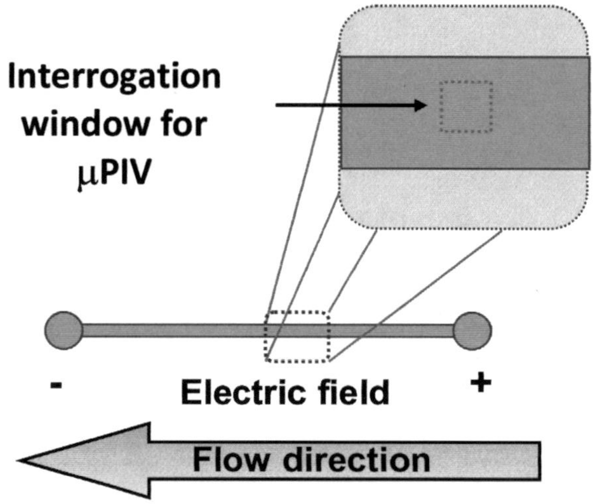

Figure 2: Experimental setup for *µPIV* measurements.

The experiments described above were performed in order to identify the best operating conditions for dielectrophoretic trapping of particles, *i.e.*, lowest electrokinetic force which benefits dielectrophoretic concentration of particles. To corroborate the results found with *PIV*, a set of iDEP experiments was carried out using a microchannel containing an array of cylindrical insulating posts. The microchannel used in this work was 1-cm-long, 1-mm-wide, 10-µm deep, and had an array of 8 columns x 4 rows of cylindrical insulating posts 200-µm in diameter and arranged 250-µm center-to-center (Figure 3).

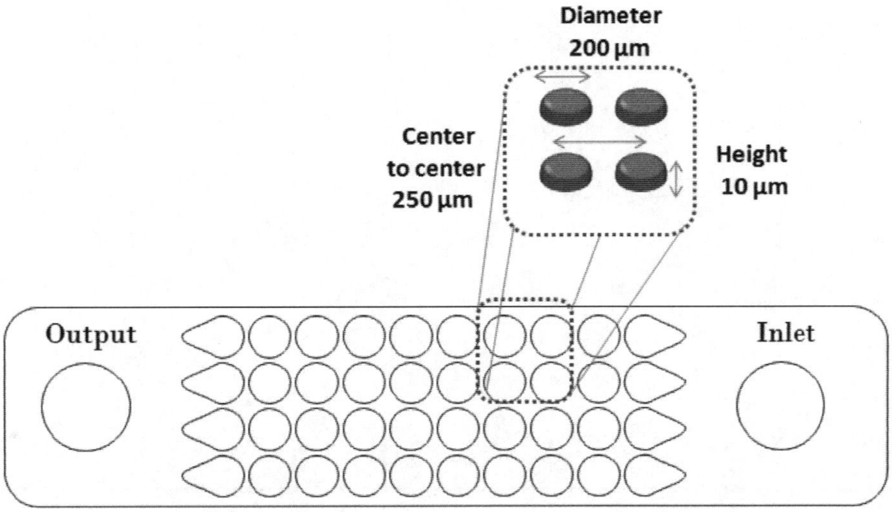

Figure 3: Microchannel for iDEP trapping.

Prior to experimentation, the microdevice was cleaned with bidistillated water and filled with the suspending medium of interest. A sample of 80 µl of the microspheres was introduced at the inlet reservoir. Care was taken in eliminating any pressure driven flow present. The experimental setup is shown in Figure 4.

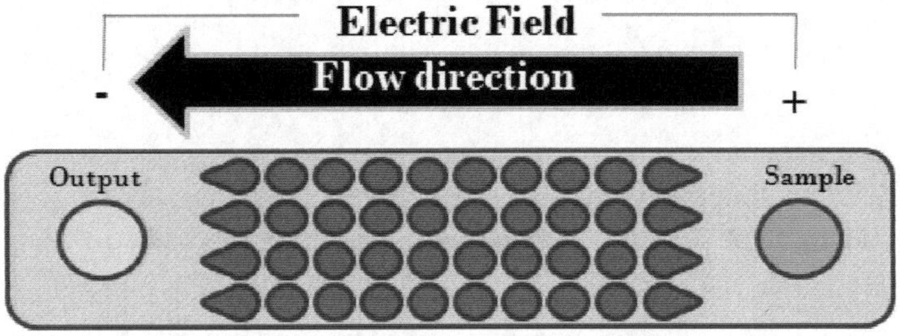

Figure 4: Experimental setup for iDEP trapping.

Results and discussion

Figures 5 and 6 show the results obtained for microparticle velocity and electrokinetic mobility, respectively, as a function of the electric field and suspending medium pH and conductivity. Electrokinetic mobility values were obtained employing Equation 4. The values employed in these calculations were: ε_m= 78, ε_0=8.85419x10^{-12} F/m y η =0.001(N·s/m^2). As mentioned, each experiment was repeated twice, the maximum standard deviation obtained with these experimental data was 15.86 %.

From Figures 5 and 6 one can observe that increasing the conductivity of the suspending medium decreases the magnitude of the EOF (lower velocity and mobility). The relation between conductivity and electroosmotic mobility can be explained considering the existence of the EDL (electrical double layer). When the conductivity of the suspending medium increases the thickness of the EDL decreases, isolating the charge on the channel surface, by a shielding effect, resulting in electroosmotic flow being diminished (Kirby *et al.*, 2005). Therefore, increasing conductivity decrease electroosmotic flow and promotes dielectrophoretic trapping of particles at lower electric field values.

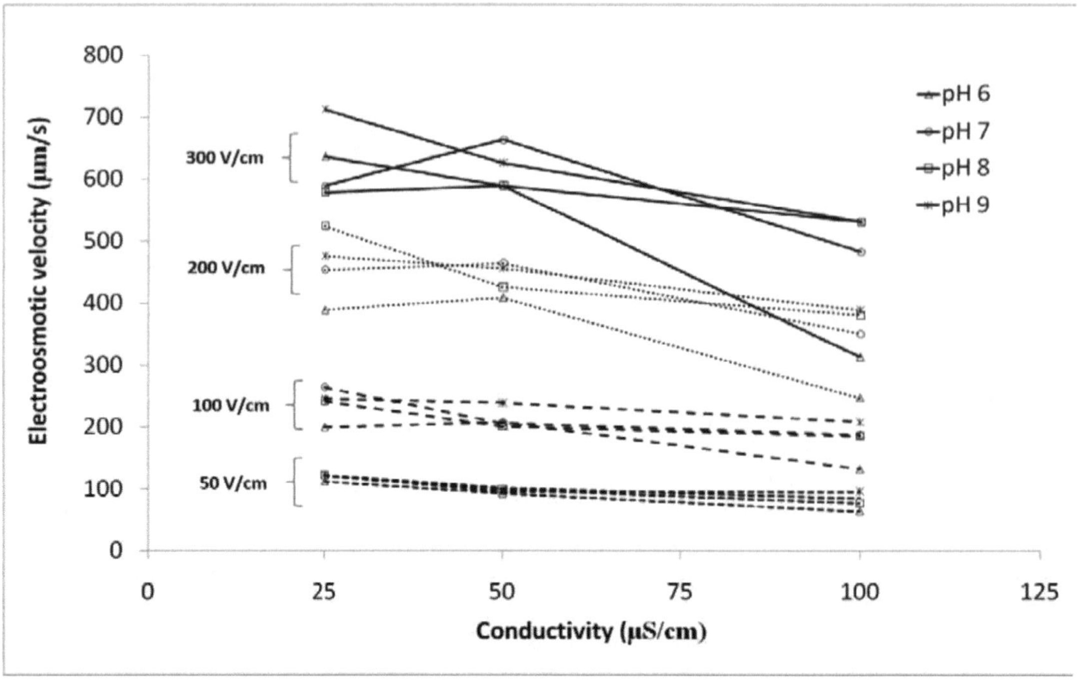

Figure 5: Microparticle velocity as a function of the electric field, pH and conductivity of the suspending medium.

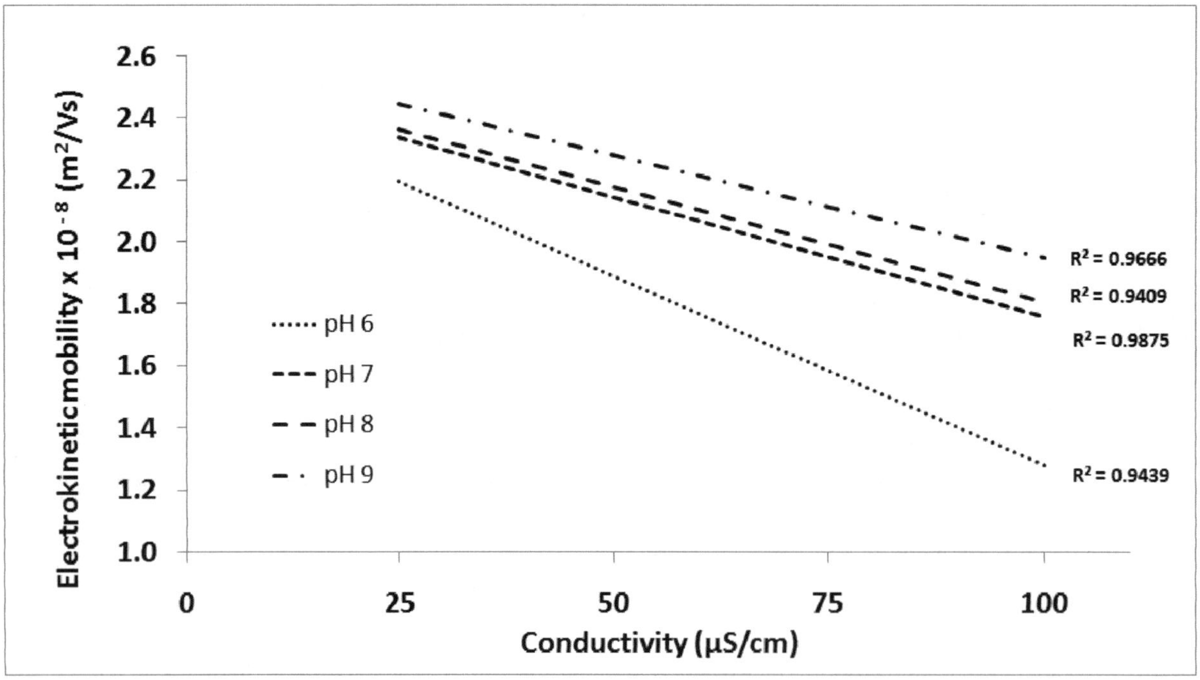

Figure 6: Electrokinetic mobility as function of pH and conductivity of the suspending medium.

In contrast, pH and electrokinetic mobility are directly related, an increasing in the pH value reflects the raise in the OH^- groups in the medium; as a result, there is an increase in the number surface silanol groups on the glass channel surface, increasing the surface charge, the zeta potential and therefore the magnitude of the electroosmotic flow (Hayes *et al*., 1993). As the electroosmotic flow increases, the energy needed to trap particles in iDEP microdevice will also rise; therefore the lower pH values will offer lower electrokinetic mobility and better dielectrophoretic trapping conditions.

These results were corroborated by carrying out iDEP experiments and are shown in Figure 7. As it can be observed, under the same electric field (E= 600 V/cm), the strongest dielectrophoretic trapping of the microparticles was obtained with a suspending medium of low pH and high conductivity (Figure 7a, pH=6 σ=100 μS/cm). Dielectrophoretic trapping decreased by increasing the pH to a value of 9 (Figure 7b: pH 9 σ=100 μS/cm) or by decreasing the conductivity (Figure 7c. pH 6 σ=25 μS/cm). The results shown in Figure 6a confirm that dielectrophoretic trapping is improved when the electrokinetic force is decreased by manipulating the suspending medium properties. These results demonstrate that operating conditions can be varied in order to enhance dielectrophoretic concentration of particles.

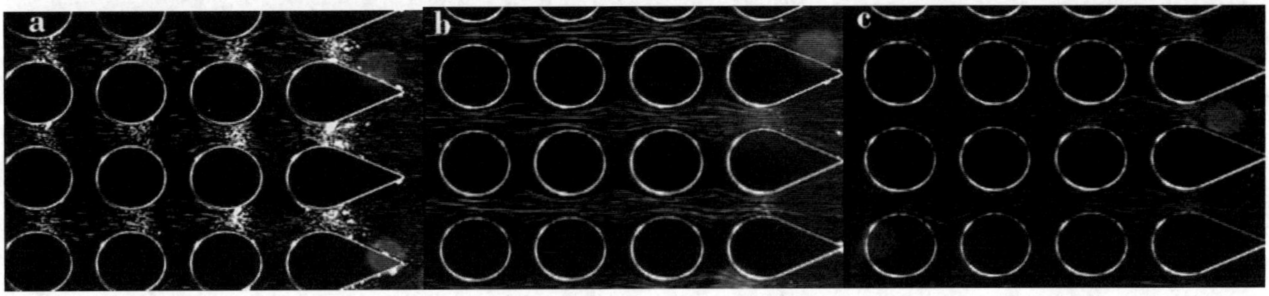

Figure 7: Effect of pH and conductivity in the dielectrophoretic trapping of microparticles at E=600 V/cm. Flow direction is from right to left, post diameter is 200 μm.

Conclusions

Dielectrophoresis is a proved technique with great potential for concentrating a wide array of bioparticles. In this work it has been demonstrated that the optimal values for dielectrophoretic trapping are low pH and high conductivity. Conductivity and pH of the suspending medium are directly related to the overall performance of the insulator-based dielectrophoretic trapping. Increasing the conductivity will produce the reduction of the electrokinetic force that prevents dielectrophoresis to occur. On the contrary, higher pH will produce stronger electroosmotic flow, resulting in a decrease of dielectrophoretic trapping of particles. These results provide with guidelines for the operation and optimization of microdevices for insulator-based dielectrophoresis.

Acknowledgments

The authors wish to thank the financial support given by the CONACYT GRANT No. CONACYT-CB-2006-53603, and *Cátedras de Investigación* of *Tecnológico de Monterrey* (CAT161 and CAT041). The authors are thankful M.Sc. Sandra Ozuna Chacón for providing support with experimental work.

References

1. Adrian R.J. (1984), "Scattering particle characteristics and their effect on pulsed laser measurements of fluid flow: speckle velocimetry vs. particle image velocimetry", *Appl. Opt.,* 23**,** 1690-91.
2. Hayes, M.A., Kheterpal, I., and Ewing, A. (1993), "Effects of Buffer pH on Electroosmotic Flow Control by an Applied Radial Voltage for Capillary Zone Electrophoresis," *Anal. Chem.*, 65:27-31.
3. Kirby, B.J, and Hasselbrink, E.F. (2004), "Zeta potential of microfluidic substrates: 1. Theory, experimental techniques, and effects on separations," *Electrophoresis*, 25:187-202.
4. Kuhn, R. and Hoffstetter-Kuhn (1993), Capillary Electrophoresis: principles and practice. 5-33.
5. Lapizco-Encinas, B.H., Simmons, B.A., Cummings, E.B., Fintschenko, Y. (2004b), "Insulator-Based Dielectrophoresis for the Selective Concentration and Separation of Live Bacteria in Water," *Electrophoresis*, 25:1695-1704.
6. Lapizco-Encinas, B.H., Ozuna-Chacón S., Rito-Palomares, M., (2008) "Protein manipulation with insulator-based dielectrophoresis and direct current electric fields. In press, *J. Chrom. A.*, 1206:45-51.
7. Lapizco-Encinas, B.H., Simmons, B.A., Cummings, E.B. and Fintschenko, Y. (2004a), "Dielectrophoretic Concentration and Separation of Live and Dead Bacteria in an Array of Insulators," *Anal.Chem.,* 76:1571-1579.
8. Li, H., and Bashir, R. (2002),"Dielectrophoretic separation and manipulation of live and heat-treated cells of Listeria on microfabricated devices with interdigitated electrodes", *Sens. Actuators, B,* 86(2-3): 215-221.
9. Markx, G.H., Y. Huang, X.F. Zhou, and Pethig R. (1994), "Dielectrophoretic characterization and separation of micro-organisms", *Microbiology*, 32:29-37.
10. Ozuna-Chacon, S., Lapizco-Encinas, B.H., Rito-Palomares, M., and Martínez-Chapa., S.O. (2008), "Performance characterization of an insulator-based dielectrophoretic microdevice", *Electrophoresis,* 29:3115-3122.
11. Prasad A.K., (2000) "Particle image velocimetry", *Curr. Sci.* 79:51-60.
12. Raffel, M., Willert, C. and Kompenhans J., Particle Image velocimetry: A Practical Guide. Berlin: Springer, 1998.
13. Rousselet, J. Markx, G. and Pethig, R. (1998),"Separation of erythrocytes and latex beads by dielectrophoretic levitation and hyperlayer field – flow fractionation.", *Colloids Surf., A*, 140: 209-216
14. Washizu, M., Kurosawa, O., Arai, I., Suzuki, S., and Shimamoto, N. (1995), "Applications of electrostatic stretch-and-positioning of DNA," *IEEE Trans. Ind. Appl.,* 31:447-456.
15. Westerweel, J. (1997), "Fundamentals of digital particle image velocimetry", *Meas. Sci. Technol.,* 8:1379-1392.

518b Thermoresponsive Microparticle Gels for Electrophoresis: PNIPAm Templated Page

Jeffery W. Thompson, Holly A Stretz, and Pedro E. Arce. Chemical Engineering, Tennessee Technological University, Tennessee Technological University Department of Chemical Engineering, PO Box 5013, Cookeville, TN 37083

Gel electrophoresis is a widely used technique for the separation of biomacromolecules, including DNA, proteins, antibiotics and others. Clinical studies, diagnostics, and lab production are some excellent examples where electrophoresis seems to work well for a large family of these molecules. However, for cases where the physical or transport properties, i.e. electrophoretic mobility and/or diffusivity are similar, as well as biomolevualr sizes are of the same order, the separation becomes very challenging. The fibrous gel morphology exposes the biomacromolecules to a "differentiating" material allowing the physical properties mentioned above to yield a motion that is different for biomacromolecules with different transport properties.

Therefore, controlling the nature and sizes of this architecture or morphology may have a beneficial impact on the separation efficiency. Rill et. al. (1995) proposed a modified architecture by templating gels with DNA (or surfactant) templates that were removed from the gels once the polymerization took place, by electrophoresis action. The resulting material, hypothesized as a "dual" porous structure, displaying both small (of the order of the inter-fiber voids) and larger (of the order of the templating agent) pores or voids, led to an improved separation efficiency.

Templating agents in gels, as mentioned above, are not the only approach to modifying gel morphology Embedded nanoparticles of varying properties are another option because of the multitude of potential alternatives that they offer regarding the physical properties of the gel. For example, the presence of nanoparticles within the gel has the potential to modify the electrokinetic properties of the gel; therefore, these nanoparticles may influence electro-osmotic flows, as it has been shown in preliminary studies for sensor applications (Matos et. al., 2006). Sedrick et. al. (AIChE/ACS, 2008) have synthesized these types of gels by using laponite nanoparticles of a given size and properties. TEM images show an interesting and excellent potential structure that may lead to an improved separation efficiency based on the modified architecture of the nano-composite gels.

Nanocomposite gels whose morphology is sensitive to temperature for drug delivery and bioseparations of proteins or DNA holds great potential. These materials feature, for example, nano or microparticles embedded in the gel structure that creates a thermo-sensitive and composite polymer with different and unique transport properties. The synthesis and characterization of thermally responsive particles as well as including them in gel matrices are described. The particles are synthesized with a precipitation polymerization crosslinking reaction and inserted into gels. In addition, electrophoresis runs are used to test the new gels in proteins motion. Both UV and visual characterization are used to determine and compare the transport (i.e. mobility and dispersion) characteristics of the new gels with standard gels in the electrophoresis runs. The new nanocomposite gels offer excellent potential to improve separation. Details for the current and future work will be offered.

Investigation of Changes In the Surface Chemistry and Morphology of Platinum Microelectrodes Subjected to a Dielectrophoretic Field

Aytug Gencoglu[1], Emily F. Cotten[1], S. Anell Pullen[1], Sarah Thompson[2], B. Selin Tosun[3], Adrienne R. Minerick[1]

1: Medical Microdevice Engineering Research Laboratory, Swalm School of Chemical Engineering, Mississippi State University, Box 9595, 39762, Mississippi State, MS
2: Mississippi School for Math and Science, Mississippi University for Women, 1100 College St, 39701, Columbus, MS
3: Istanbul Technical University, Department of Metallurgical and Materials Engineering, ITU Metallurgical & Materials Engineering Department, 34469, Maslak, Istanbul, Turkey

Platinum microelectrodes are widely used in microdevice applications as they are considered to be inert and noncorrodible. Chemical and morphological changes that occur on platinum surfaces are well documented, and results from our laboratory suggested that platinum microelectrodes subjected to electric fields undergo changes that affect their performance. Chemical and morphological changes on platinum microelectrodes subjected to ≤1.5 V_{pp} and 6 V_{pp} electric fields in NaCl and KH_2PO_4/K_2HPO_4 solutions were investigated with cyclic voltammetry and scanning electron microscopy. The results show that a number of chemical and morphological changes occur on platinum surfaces over time, which may significantly affect reusability of microdevices utilizing them.

Introduction

Platinum is a popular electrode material in biological microdevice applications because it is considered to be inert and noncorrodible. As a result, platinum microelectrodes are often used to generate AC and DC electric fields in microdevices, and any electrochemical effects are largely ignored in microanalytical systems. This is a valid assumption over a limited range of voltages; platinum does not dissolve in electrolytic solutions when the applied potential is between the potentials of hydrogen and oxygen evolution [1,2]. A large number of microdevices employ electrophoretic techniques such as capillary electrophoresis (CE), and capillary zone electrophoresis [3]. Electric fields are used to facilitate electroosmotic flow in microdevices [3-5]. Electrode chemistry effects are undesirable in these applications because they rely on either AC or DC electric field effects for separation of cells and particles [3,6-10]. However, in a limited number of applications, such as isoelectric focusing, pH gradients are employed to separate cells and particles [3,11]. These pH gradients can be generated in part by hydrolysis reactions at the microelectrodes, whereby surface changes on platinum electrodes are undesirable, since they can affect the hydrolysis reactions [11]. AC fields have been touted as advantageous over DC fields on the grounds that undesirable reactions at electrode surfaces can be avoided with AC fields. This work demonstrates the fallacy of this assumption.

Chemical and physical changes that platinum electrodes can undergo in AC and DC electric fields are well documented in the electrochemistry literature and fall into five categories, namely, chemical reactions of platinum, oxide formation, dissolution of platinum, adsorption on platinum surfaces and Ostwald ripening. These mechanisms are discussed in more detail in our full paper [12].

The presence of these chemical and physical changes were investigated in conditions similar to those in currently developed microdevices. Recently in our group, platinum microelectrodes were utilized to disrupt red blood cell membranes by the application of AC electric fields; it was observed that older platinum microelectrodes ruptured the red blood cells more rapidly than newer ones [13]. This observation suggests that some properties of platinum microelectrodes change with time when they are subjected to AC electric fields, which in turn impacts electric fields in microdevices. Property changes in platinum microelectrodes would have implications for analytical microdevices, because changes in the electric field can affect the behavior of cells by changing the speed of electrophoretic separations or electric field gradients. It is desirable to understand the chemical and morphological changes that platinum microelectrodes undergo and determine the conditions (frequency, potential, etc.) under which microdevice electric fields remain reliably constant.

To this end, cyclic voltammetry (CV) has been used to determine the chemical changes that occur on platinum microelectrodes in NaCl and K_2HPO_4/KH_2PO_4 solutions. These electrolytes are the constituents of phosphate buffer saline (PBS) solution, which is commonly used in medical microdevices. The solutions used in the CV study had the same ionic strength as the PBS solution used in other studies in our lab, which mimics the salinity of human blood. CV plots obtained with NaCl and K_2HPO_4/KH_2PO_4 solutions were compared to determine the effect of chlorine on chemical and morphological changes on platinum surfaces under the experimental conditions. Morphologies of new platinum microelectrodes and those subjected to electric fields were compared by scanning electrochemical microscopy (SEM).

A more complete surface chemical analysis of platinum microelectrodes under AC and DC conditions can be found in our full paper, "Morphological and Chemical Changes on Platinum Microelectrode Surfaces in AC and DC Fields with Biological Buffer Solutions" [12].

Materials and Methods

Cyclic voltammetry experiments were carried out using new microelectrodes made of 0.1 mm diameter platinum wire with 99.99% purity (Goodfellow, Huntingdon, UK) as the working electrode, a 1 mm diameter platinum wire as the auxiliary electrode, and a 1 mm diameter silver wire as the quasi-reference electrode (QRE). Unbiased potential signals of 1.5 V_{pp} and 6 V_{pp} were applied, with a scan rate of 80 mV/s. CV investigation of new platinum working electrodes was carried out in NaCl solution (0.274 M NaCl) and potassium phosphate solution (PPS) (0.036 M KH_2PO_4, 0.012 M K_2HPO_4) in order to observe any reactions involving ions present in PBS solution separately.

To study the morphological changes on platinum microelectrodes using SEM, two new microelectrodes were subjected to a 2 kHz AC field in freshly prepared NaCl solution and two were subjected to a 2 kHz AC field in freshly prepared PPS solution for 5 hours. For both solutions, one of the prepared microelectrodes was subjected to an AC field potential of 1 V_{pp} and the other to a field potential of 6 V_{pp} with an electrode spacing of 175 μm for a resulting field strength of 3.43E4 V/m. All four microelectrodes were prepared with new platinum wire and fresh electrolyte solution.

SEM images were taken using a JEOL JSM-6500F field emission microscope (JEOL USA, Peabody, MA) under a vacuum of 2.72E-6 torr. All samples were energized with a 5.0

keV electron beam. The portion of surface covered by a feature was estimated by laying a 5 μm x 5 μm grid over the SEM images and comparing the number of vertices on a feature to the total number of vertices.

Results and Discussion

Cyclic voltammetry plots of platinum microelectrodes in NaCl and potassium phosphate solutions with 1.5 V_{pp} and 6 V_{pp} signals are shown in Figure 1. The surface areas of the microelectrodes were not controlled; therefore, a normalization of currents according to surface area and quantitative comparison of results were not possible. Potentials are given as "volts against silver quasi-reference electrode" (V_{AgQRE}). Only the sections from -1.25 V_{AgQRE} to 1.25 V_{AgQRE} of CV plots generated by a 6 V_{pp} signal are shown, since the currents abruptly became excessively high beyond these values. This sharp increase in the current suggests a large capacitative current, which is caused by the restructuring of the electrolyte double layer near the electrode surface and does not indicate electron transfer to or from the electrode surface. The negative sweeps revealed electrochemical oxidation peaks at -0.15 V_{AgQRE} with 1.5 V_{pp} signals and at -0.25 V_{AgQRE} with 6V_{pp} signals, while the positive sweep featured a plateau, indicating that the oxidation of platinum microelectrodes is

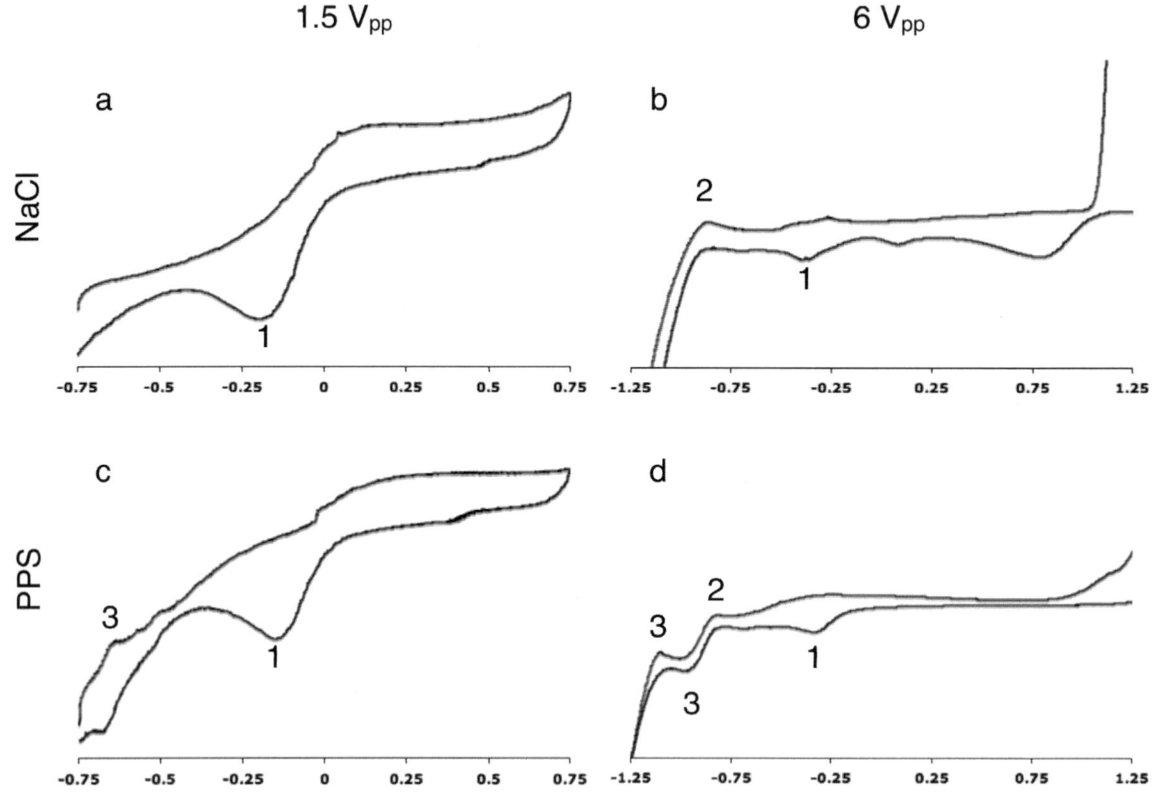

Figure 1: Cyclic voltammetry plots of platinum microelectrodes with a platinum counter electrode and a silver reference electrode at a scan rate of 80 mV/s. a) NaCl solution with a 1.5 V_{pp} signal, b) NaCl solution with a 6 V_{pp} signal, c) Potassium phosphate solution with a 1.5 V_{pp} signal, d) Potassium phosphate solution with a 6 V_{pp} signal. Features labeled on the figure indicate: 1) Electrochemical oxidation of platinum, 2) Hydrogen formation from water reduction, 3) Potassium adsorption/desorption.

irreversible. Hydrogen formation from water reduction was observed at -0.80 V_{AgQRE} with $6V_{pp}$ signals but not with 1.5 V_{pp} signals, as was expected. The current increased as the potential increased with 1.5 V_{pp} signals, suggesting an adsorption process. These results are in agreement in with a published CV investigation of platinum foil carried out in a PBS solution prepared with Na_2HPO_4, NaH_2PO_4 and NaCl, and reporting chloride and oxygen formation at higher positive potentials [14]. In the CV plots generated in our lab with potassium phosphate solution, a peak in the positive sweep was observed at a potential of -0.65 V_{AgQRE} with a 1.5 V_{pp} signal and at -1.08 V_{AgQRE} with $6V_{pp}$ a signal, suggesting potassium adsorption on the platinum microelectrode surfaces. A corresponding peak was also observed in the negative sweep at -0.97 V_{AgQRE} with a 6 V_{pp} signal. These peaks were not present in CV plots generated with NaCl solution or in the CV plots generated with a potassium-free buffer presented in another study [14]. In conclusion, CV verified that oxide formation, hydrogen formation from water reduction, chloride formation and potassium adsorption take place on the platinum microelectrode surfaces.

SEM was used to determine changes in the surface morphology of platinum microelectrodes after 5 hours in a 2 kHz, 1 V_{pp} or 6 V_{pp} AC electric field. It was found that the morphology of platinum microelectrodes changed considerably after 5 hours in an AC electric field. A comparison of new platinum microelectrodes and platinum microelectrodes

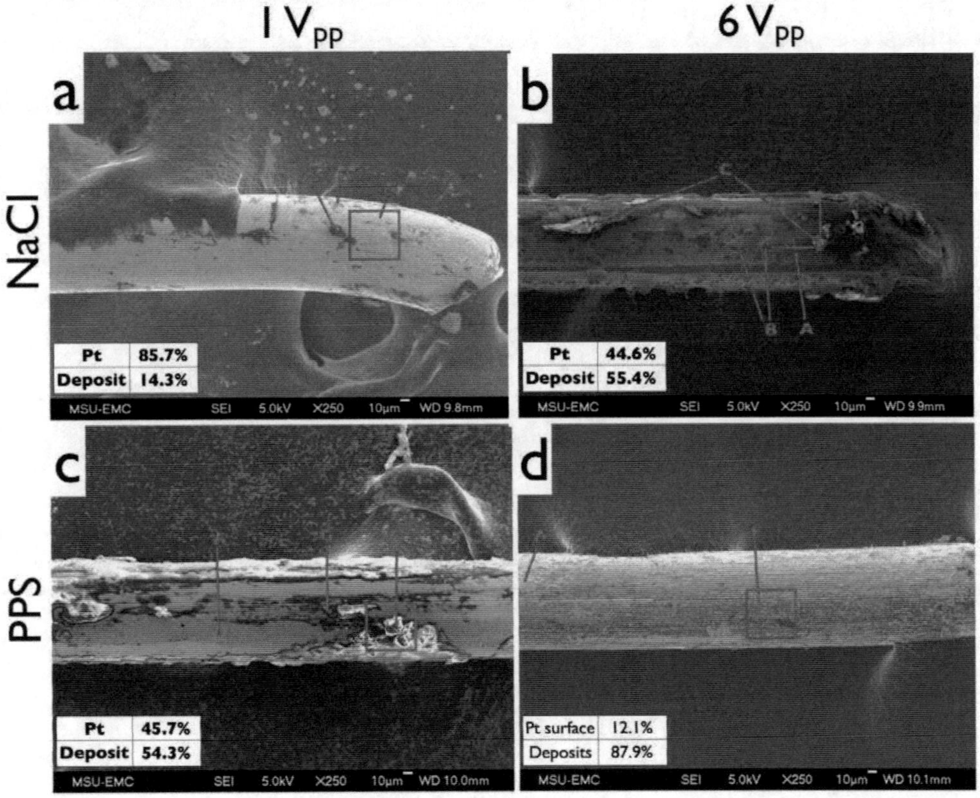

Figure 2: Scanning electron microscopy images of platinum microelectrodes at 250X magnification. Values in the lower left corner of each image give the percentage of surface area of morphological features, estimated by a 5 µm x 5 µm grid over the whole surface. a) NaCl solution with 1 Vpp signal, b) NaCl solution with 6 Vpp signal, c) Potassium phosphate solution with 1 Vpp signal, d) Potassium phosphate solution with 6 Vpp signal. Red squares indicate areas magnified in Figure 3. Red letters indicate features described in the text.

that were subjected to 1 V_{pp} and 6 V_{pp} electric fields in NaCl and potassium phosphate solutions for 5 hours is given in Figure 2, along with the percentages of surfaces that are affected by morphological changes, as estimated by application of a 5 μm x 5 μm grid over the images. Platinum surfaces affected by dissolution of platinum but otherwise free of deposits were regarded as free of morphological features in the estimation of these percentages.

The most notable features that appeared after prolonged exposure to the electric field were deposited material on the electrode surface and platinum dissolution. This effect is most pronounced in the platinum microelectrode subjected to a 6 V_{pp} electric field in NaCl solution. On this microelectrode, it was observed that large areas of the electrode surface were dissolved (Marked as A on Figure 2a), as opposed to the more localized deformations due to dissolution that were observed with the other electrode conditions. In addition, two different material deposits were observed, one consisting of large particles (Marked as B on Figure 2a) and the other consisting of layers of crystalline deposits (Marked as C on Figure 2a), the latter of which can be seen in higher magnification in Figure 3a. Material deposition (Marked as D on Fig 2b) and surface changes (Marked as E on Figure 2b) were observed on 14.3% of the microelectrode subjected to a 1 V_{pp} electric field in NaCl solution. Part of the surface of this microelectrode was obscured by material from the mounting tape and was excluded from analysis. Platinum dissolution was much less severe on the microelectrode which was subjected to a 6 V_{pp} electric field in potassium phosphate solution compared to the one subjected to the same field in NaCl solution, but 87.9% of the surface was covered with deposited material (Marked as F on Figure 2c). Surface changes (Marked as G on Figure 2d) and large areas of material deposits (Marked as H on Figure 2d) covering 54.3%

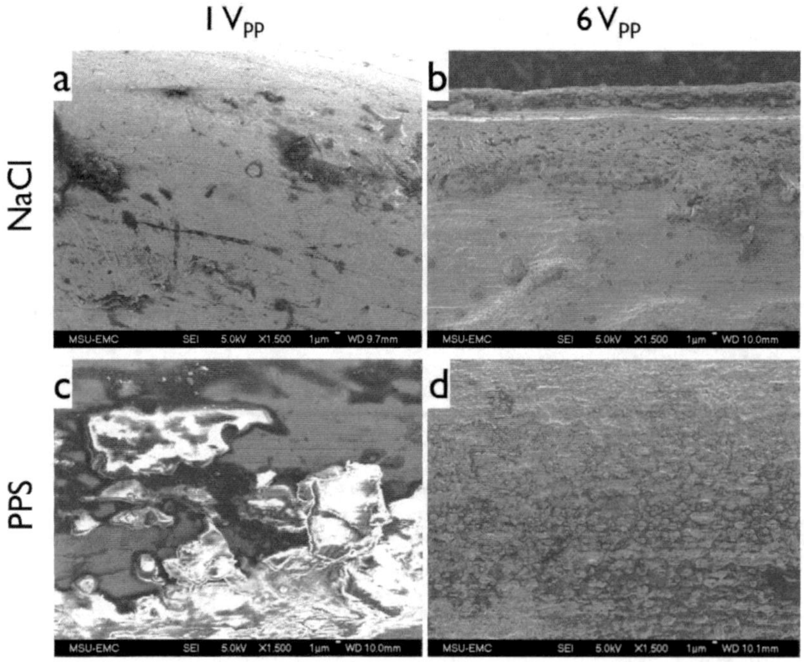

Figure 3: SEM images of red boxed features on Figure 2 at X1500 magnification. a) NaCl solution with 1 Vpp signal, b) NaCl solution with 6 Vpp signal, c) Potassium phosphate solution with 1 Vpp signal, d) Potassium phosphate solution with 6 Vpp signal.

of the microelectrode surface were observed on the microelectrode subjected to a 1 V_{pp} electric field in potassium phosphate solution. Further chemical analysis by energy dispersive spectroscopy (EDS) is presented in our full paper [12].

Images of the observed features taken at higher magnifications are shown in Figure 3. In Figure 3a, it can be seen that the platinum surface of the microelectrode subjected to a 6 V_{pp} field in NaCl solution has a rough topography compared to the smooth appearance of the other microelectrodes, further demonstrating the severity of platinum dissolution on this sample. It was observed from Figure 3a and Figure 4a and another SEM image of the sample shown in Figure 3a that in a 6 V_{pp} electric field in NaCl solution, platinum dissolution is severe enough that the overall thickness of the wire is reduced by ~10% over the 5 hour experiment. Dissolution of platinum is observed in all cases, and it is known that dissolved platinum can be redeposited on the platinum surface when the electric field is reversed, as occurs in AC applications, causing heterogenous growth of platinum crystals on the surface [15].

Conclusions

Chemical and morphological changes to platinum microelectrode surfaces subjected to 1 V_{pp}/175 µm and 6 V_{pp}/175 µm AC electric fields in chlorine-bearing and chlorine-free solutions were compared. These conditions are commonly encountered in medical microdevices. More drastic chemical and morphological changes to microelectrodes were observed in 6 V_{pp} fields as opposed to 1 V_{pp} fields. Surface changes and dissolution of platinum, as well as material deposition and the possible redeposition of platinum were observed in all cases. Results from our own laboratory suggest that these chemical and morphological changes affect the electric field produced by platinum microelectrodes and subsequent chemical analyses.

These findings point to the fact that despite its well known resistance to corrosion, platinum is susceptible to chemical and morphological changes which can potentially alter performance of medical microdevices over time and limit their reusability, as changes in the electric field mean changes in the cell responses and inconsistent behaviors. In microdevice design, the media that the microelectrodes will contact and the electric fields they will be subjected to must be considered when estimating reusability. Insulator-based (electrodeless) microdevices can be considered if reusability and extended lifetime are desired.

References

1. Christensen, P. A., Hamnett, A., "Techniques and Mechanisms in Electrochemistry." Blackie Academic and Professional, Glasgow:1994.
2. Ghilane, J., Guilloux-Viry, M., Lagrost, C., Simonet, J., Hapiot, P., *Journal of the American Chemical Society,* **129,** 6654-6661 (2007).
3. Suntornsuk, L., *Journal of Chromatographic Science,* **45,** 559-573 (2007).
4. Rodriguez, I., Chandrasekhar, N., *Electrophoresis,* **26,** 1114-1121 (2005).
5. Barker, S. L. R., Ross, D., Tarlov, M. J., Gaitan, M., Locascio, L. E., *Analytical Chemistry,* **72,** 5925-5929 (2000).
6. Chou, Chia-Fu, Zenhausern, F., *IEEE Engineering in Medicine and Biology Magazine,* **22,** 62-67 (2003).
7. Lynch, B. P., Hilton, A. M., Simpson, G. J., *Biophysical Journal,* **91,** 2678-2686 (2006).
8. Kang, K. H., Kang, Y., Xuan, X., Li, D., *Electrophoresis,* **27,** 694-702 (2006).

9. Barbulovic-Nad, I., Xuan, X., Lee, J. S. H., Li, D., *Lab on a Chip,* **6,** 274-279 (2006).
10. Pysher, M. D., Hayes, M. A., *Analytical Chemistry,* **79,** 4552-4557 (2007).
11. Cabrera, C. R., Finlayson, B., Yager, P., *Analytical Chemistry,* **73,** 658-666 (2001).
12. Gencoglu, A., Minerick, A. R., "Morphological and Chemical Changes on Platinum Microelectrode Surfaces in AC and DC Fields with Biological Buffer Solutions", to be submitted to *Lab on a Chip.*
13. Walton, M., (2007). *MicroMadness and Me: Erythrocyte Rupturing Responses in Dielectrophoretic MicroDevices.* Unpublished manuscript.
14. Mailley, S., Hyland, M., Mailley, P., McLaughlin, J. A., McAdams, E. T., *Bioelectrochemistry,* **63,** 359-364 (2004).
15. Vleeming, J.H., Kuster, B.F.M., Marin, G.B., Oudet, F., Courtine, P., *Journal of Catalysis,* **166,** 148-159 (1997).

518d An Electrokinetic-Hydrodynamics-Ekhd Based Approach to Determine Effective Transport Parameters: Illustrative Results and Comparison

Jennifer Anne Pascal[1], Mario Oyanader[2], and Pedro E. Arce[2]. (1) Chemical Engineering, Tennessee Tech University, PH-214, Cookeville, TN 38505, (2) Chemical Engineering, Tennessee Technological University, Department of Chemical Engineering, P.O. Box 5013, Cookeville, TN 38505

A number of contributions have been made to the field of electrokinetics, i.e. electrophoresis, EOF, etc. as it relates to the behavior of diffusion and hydrodynamics in various systems. Many of these efforts have been influenced by the work conducted by the group of Giddings (see, for example, Martin and Giddings [1], [2]) who worked in the area of field flow fractionation (FFF). In this separation technique, an orthogonal applied field (i.e. gravitational, electrical, etc.) drives the motion of the charged molecules towards the channel walls; thus, the solute that is most susceptible to the influence of the field locates closest to the walls, and, depending on the solute properties, different species reach the wall at different locations.

Among illustrative contributions and without being thorough in the literature citations, Giddings ([1]), based on nonequilibrium theory, predicted retention times in various types of FFF devices. Brenner and Edwards ([3]), examined a Couette-based flow apparatus with cross flow in which equations governing the transport of Brownian solute particles (i.e. diffusivity and velocity) were developed. More recently, Horiuchi et. al. ([4]) have examined two dimensional flow in rectangular microchannels with non constant electrostatic potential on the walls. They obtained analytical solutions for the velocity profile and pressure as well as the effective diffusivity. These "effective" parameters control the quality of the separation in the device.

A powerful mathematical framework that is currently available ([5]) for these types of problems is the coupling of EKHD with the spatial averaging approach ([6], [7], [8], [9]). This mathematically efficient approach allows for up-scaling as well as systematization from the micro (molecular or continuum) scale to the macro (bulk) scale. One of the advantages of this combined approach is the prediction of optimal separations for a wide variety of cases with a minimum effort. Moreover, the analytical results are rather simple mathematical functions of the fundamental physical parameters of the system. In this presentation, the authors will present illustrative results for both the Poiseuille as well as the Couette hydrodynamic flows (i.e. optimal separation times). Furthermore, the authors will compare these predictions with results found in the literature related to these cases and draw conclusions about the possible more suitable description and general validity of such predictions.

[1] Giddings, J. C., Science, 1993, 260, 1456.

[2] Martin, Michel; Giddings, J.C. Journal of Physical Chemistry. 1981, 85, 727.

[3] Brenner, H.; Edwards, D.A. Macrotransport Processes. Butterworth-Heinemann Series in Chemical

Engineering, 1993.

[4] Horiuchi, K. Dutta, P., Ivory, C. "Electroosmosis with Step Changes in

Zeta Potential in Microchannels." AIChE Journal 53 (2007): 2521.

[5] Oyanader, M. A. and P. Arce, Electrophoresis, 26, 2857 (2005).

[6]Slattery, J. C., Momentum, Energy and Mass Transfer in Continua, Krieger, New York, 1981.

[7] Whitaker, S., Chem. Eng. Sci. 1985, 40, 1387.

[8] Cwirko, E. H.; Carbonell, R. G., . J. Colloid Interface Sci. 1989, 129, 513.

[9] Arce, P.; M. Oyanader, and J. Pascal, "Electrokinetic Hydrodynamics: An Introductory Graduate level Course." To be submitted to Chemical Engineering Education.

518e Electrokinetic-Hydrodynamics (EKHD): An Efficient Framework for Systematic Research

Jennifer Anne Pascal[1], Mario Oyanader[2], and Pedro E. Arce[2]. (1) Chemical Engineering, Tennessee Tech University, PH-214, Cookeville, TN 38505, (2) Chemical Engineering, Tennessee Technological University, Department of Chemical Engineering, P.O. Box 5013, Cookeville, TN 38505

Currently there are many problems in Chemical Engineering that involve the application of an applied electrical field to a fibrous or porous media in a variety of relevant technological processes. Examples of these types of problems are the separation of biomacromolecules, such as proteins and DNA, bioremediation, drug delivery, and coating flows, among others. These types of problems can be described by the fundamentals of "electrokinetic-hydrodynamics" or EKHD, for short ([1]). What makes the applications within the domain of electrokinetic-hydrodynamics unique is that unlike electrochemical systems, bulk motion of the fluid occurs. Therefore, we can view EKHD as involving two domains: the motion of the fluid (electrohydrodynamics) and the motion of the solute/analyte (electro and convective-diffusive transport). It is apparent that these two domains are representative of two different scales, the continuum (fluid) and the discrete (solute) . This contribution will discuss how EKHD is an efficient framework for the investigation of systems with low and high values of applied electrical fields.

For the understanding the behavior of the systems in EKHD, one must have a basic background in fluid mechanics, and mass, momentum, and energy conservation. These concepts can then be used to link the principles of electrokinetics to those of hydrodynamics to model the fluid motion, i.e. "electrohydrodynamics". Once this knowledge is acquired, then the solute/analyte problem can be described. The coupling between these two domains (i.e. fluid and solute) can be effectively handled by using the spatial averaging technique ([2]). This powerful up-scaling approach allows for the computation of analytical expressions which lead to effective, or alternative "macro," coefficients (i.e. effective velocity and diffusivity) that represent the macroscopic behavior of the system. These coefficients can then be used to determine information relevant to practical applications such as the optimal time of separation of biomacromolecules, time for the cleaning protocol in soil remediation, among others. In this contribution, the authors will discuss the method of EKHD and the different elements associated with it; for the didactic aspects, typical assignments/exercises associated with this field of research including, problems and projects, will be included.

518f Polyacrylamide Gel Modified with Poly-N-Isopropylacrylamide Coated Gold Nanoparticles for Electrophoresis

Jyothirmai J. Simhadri, Holly A. Stretz, and Pedro E. Arce. Chemical Engineering, Tennessee Technological University, Cookeville, TN 38505

Traditionally, separation of proteins was accomplished by polyacrylamide gel electrophoresis; usually, gel structure is modified by change in concentrations of polymers in solution. This approach has worked well for many type of macromolecules; however when the electrophoretic mobility and the size of the macromolecules do no cooperate efficiently, the separation of this type of mixture become challenging. Therefore, other approaches must be identified. For example, by modifying the structure of gels by either "macro-voids" (Rills et al, 1995) or by adding nanoparticles (Thompson et al, 2008) have proven successfully in increasing separation efficiency. In this contribution, the synthesis and characterization of core/shell nanoparticles as well as dispersing them into the polyacrylamide gel matrix are described. The particles are synthesized by surfactant free emulsion polymerization and dispersed into the gels. UV-Vis spectrophotometer is used to characterize the particles. Characterization of the material structure and electrophoresis on albumin proteins is performed to confirm the effectiveness of the modified gel matrix on their separation efficiency. Comparison with other type of nanocomposite gel matrix will be also discussed.

518g Characterizing the Abundance and Activity of Soil Microbes by Capillary Electrophoresis Using Single-Strand Conformational Polymorphism

Alice C. Jernigan[1], Greg Thoma[1], Duane C. Wolf[2], and Christa N. Hestekin[1]. (1) Chemical Engineering, University of Arkansas, 3202 Bell Engineering, Fayetteville, AR 72701, (2) Crop, Soil and Enviornmental Sciences, University of Arkansas, 3202 Bell Engineering, Fayetteville, AR 72701

Bioremediation is the process by which microbes are used to break down soil contaminates. Although bioremediation has the potential to be a useful and potentially less expensive and more natural method of soil reclamation, there is limited information about the actual microbial communities that perform the remediation and how their abundance and activity changes with time. The identification and assessment of activity of microbial soil communities can be accomplished by isolating DNA and RNA from soil samples. These DNA/RNA samples can then be characterized by a technique called single-strand conformational polymorphism (SSCP) analysis of the 16S gene. The 16S gene is highly conserved among microorganisms, but has unique small variable regions that can be used to differentiate between species. These regions can be amplified by polymerase chain reaction (PCR) giving small fragments of the gene sequence. SSCP, a method that is normally used to detect mutations, is the heat denaturing of these fragments resulting in single strand DNA. The refolding of the single strand DNA will form different secondary structures (conformations) depending on the nucleotide sequence. These different conformations can be separated due to their different electrophoretic mobility, even if they are the same size. The traditional method for looking at bioremediation is by viable plate counts or microscopic examination. These methods are both time consuming and can underestimate the numbers due to the inability to culture many soil organisms. The increased speed and automation of capillary electrophoresis (CE)-SSCP will allow for more rapid and reliable results. By isolating both DNA and RNA from the same soil sample, at the same time, a more accurate picture of the microbial community in those samples will be available. This will be important in the study of contaminated soils and the monitoring the progress of bioremediation.

580a High Resolution In Situ Temperature Measurement In Microfluidic Systems Using Brownian Motion of Nanoparticles

Kwanghun Chung[1], Jae Kyu Cho[2], Victor Breedveld[1], and **Hang Lu**[3]. (1) Chemical and Biomolecular Engineering, Georgia Institute of Technology, 311 Ferst Dr. NW, Atlanta, GA 30332, (2) Chemical & Biomolecular Engineering, Georgia Institute of Technology, 311 Ferst Dr. NW, Atlanta, GA 30332-0100, (3) School of Chemical and Biomolecular Engineering, Georgia Institute of Technology, 311 Ferst Dr. NW, Atlanta, GA 30332-0100

Miniaturization and integration of functional components on-chip offer great advantages. For many applications, the ability to control and measure temperature inside of microfluidic devices while in operation is critical as temperature affects biological processes and temperature variation perturbs such processes, e.g. embryo development. In addition, temperature variation can be a side-effect of some microfluidic techniques, e.g. Joule heating leading to hyperthermic cell damage. Thus, a reliable in situ temperature measurement method is required to design and operate microsystems effectively. Spectroscopic methods for temperature measurement in microfluidic devices have been developed, taking advantage of temperature-dependent properties of chemicals such as Rhodamine-B. However, there is still a need for both a simple inexpensive in situ measurement and with biological samples without toxicity effect from these dyes. We present for the first time 3D temperature measurements in microfluidic systems using Brownian motion of nanoparticles in the presence of live biological samples. The technique takes advantage of well defined thermal correlation of Brownian motion. Because the kinetic energy of small particles and fluid viscosity are functions of temperature, by measuring the Brownian diffusivity of suspended particles of known sizes, the surrounding liquid temperature can be determined. We measured mean square displacements (MSD) of 500-nm fluorescent particles for temperatures ranging from 5 °C to 50°C; the calculated temperature shows good agreement with the temperature measured by a thermocouple. Since tracking nanopaticles does not rely on fluorescence intensity, MSD measurements are not affected by the power of the incident light, spatial variation of excitation, concentration artifacts, and photobleaching. Therefore, this method is superior in its reproducibility and reduced systematic errors. To demonstrate the capability of this method as a 3D temperature mapping tool, we obtained Z-stack images of nanoparticles in a temperature controlled microfluidic channel. Nanoparticles out of focus can be easily discarded by MSD analysis algorithm, thereby achieving optical sectioning without expensive multiple-photon scanning setup. We showed temperature measurement with a resolution of ~ 4 ìm in the z-axis. The mapped 3D temperature profile in the channel was in good agreement with numerical simulations in COMSOL. Finally, we used this method for temperature measurement in the presence of biological samples.

580b Micro-Device to Detect Bacterial Proliferation Using Reactance Measurements at Multiple Frequencies

Sachidevi Puttaswamy and Shramik Sengupta. Biological Engineering, University of Missouri, 1406 E Rollins Rd., 254 AEB, Columbia, MO 65201

Microorganisms are associated with the food we eat and with various kinds of animal and human diseases. Often after the initial infection, the concentration of bacteria in our sample of interest (fruit juice, milk, blood etc) is low (≤100 CFU/ml). Being able to detect them quickly at low concentrations would enable us to take preventive action faster and avert harmful effects. Classical techniques for detecting bacteria (plate culture) are slow, tedious and labor intensive. Automated approaches (biosensors and microfluidic systems) are not able to detect target pathogens quickly at low concentrations - primarily because of low volume fraction of the pathogens and/or the time they take to diffuse to a sensor surface. Our microfluidic detection system provides a means to meet this "real world" need - of being able to rapidly (<4 hours) detect the presence of low concentrations (<100 CFU/ml) of bacteria in samples such as fruit juice, milk and blood.

The technique relies on the fact that bacteria can store large amount of charge and thus increase the bulk capacitance of the liquid in which they are suspended. An increase in the number of bacteria results in an increase in the bulk capacitance of the suspension. Taking measurements at multiple frequencies from 40-100 MHz enables us to accurately measure the bulk capacitance of a suspension - and hence detect bacterial proliferations at low levels of infestations within estimated 3-4 cycles of division (~1.5 to 2 hours for E coli).

580c Optimization of Micro-Fluidic Network Geometries for Micromosaic Immunoassays

Nicholas S. Lynn[1], Brian Murphy[2], Charles S. Henry[2], and David S. Dandy[1]. (1) Department of Chemical and Biological Engineering, Colorado State University, 100 Glover building, Fort Collins, CO 80523, (2) Department of Chemistry, Colorado State University, Fort Collins, CO 80523

Protein patterning is ubiquitous in the creation of sensing motifs that rely on receptor-ligand binding for selectivity. Autonomous micro-fluidic networks (μFNs) have the potential to greatly aid in the development of simple, robust methods for deposition of biological capture agents onto a solid substrate. Unfortunately, μFN-based micro-immunoassays involving multiple binding regions suffer from the formation of an analyte depletion layer adjacent to the micro-channel floor where binding occurs. As a result, the time required to achieve equilibrium binding states for all binding surfaces is strongly dependent on both the flow conditions within the micro-channel and the geometries of the channel itself. These systems exhibit a critical flow rate (characterized by the linear velocity) below which the system becomes diffusion limited and assay times grow well beyond practical limits. For optimal μFN assay times that utilize minimal sample volumes, there is a delicate balance between the absolute flow rate through the system and the overall time in which that flow rate can be maintained. Here we report a simple method to control the flow conditions in autonomous μFNs by manipulation of simple geometric and operational parameters in order to both minimize assay time and required sample volume. The system is shown to be very flexible, providing precise control over the flow conditions over a wide rage of flow rates, utilizing only a micropipette and simple soft lithographic methods.

580d Automated Monocyte Depletion for CD4 Counting at Resource Limited Settings

Xuanhong Cheng, Materials Science and Engineering/Bioengineering, Lehigh University, 5 E. Packer Ave, Whitaker Lab, Bethlehem, PA 18015, Mehmet Toner, Massachusetts General Hospital/Shriners Burn Hospital/Harvard Medical School, 51 Blossom Street, Boston, MA 02114, and William Rodriguez, Massachusetts General Hospital/Partners AIDS Research Center, 149 13th Street, Charlestown, MA 02125.

Background: Practical HIV diagnostics are urgently needed in resource-limited settings. Accurate CD4+ T cell count is a critical clinical indicator for HIV disease staging and treatment monitoring. To address the limitations of flow cytometry, we recently developed a simple CD4 counting device based on immunoaffinity cell isolation in a single microfluidic channel functionalized with anti-CD4 antibodies. By isolating target cells from 10 uL of unprocessed whole blood, we could distinguish clinically relevant thresholds of 200, 350, and 500 cells/uL with high sensitivity (>85%) and specificity (>94%). However, this device is incapable of exact CD4 counting, especially for CD4 counts <200 cells/uL, due to monocyte contamination. To improve its accuracy, we added a monocyte depletion chamber upstream of the CD4 capture channel. Performance of the cascaded device is presented in this work.

Methods: We use microfluidic cell affinity chromatography to first deplete monocytes and then specifically isolate CD4+ T cells with high efficiency directly from 10 uL of unprocessed whole blood. CD4 counts are obtained afterwards under an optical microscope in a rapid, simple and label-free fashion. We compared CD4 counts from dual-chamber (a monocyte depletion chamber cascaded with a CD4 T cell capture chamber) and single-channel (with anti-CD4 alone) devices and conventional flow cytometry among HIV+ subjects over a wide range of absolute CD4 counts. We analyzed data via Bland-Altman comparison.

Results: Compared to the single channel devices, CD4 counts determined in the cascaded devices better correlate with the measurements by flow cytometry (R2=0.90 in dual chamber vs. R2=0.77 in single chamber devices). A Bland-Altman comparison demonstrates a bias of +69 cells/uL for the single chamber device, with limits of agreements of -66 to +204 cells/uL. Double chamber devices show a much closer agreement with flow cytometry, with a bias of +12 cells/uL and limits of agreements of -55 to +79 cells/uL respectively.

Conclusions: The dual chamber microfluidic device with a monocyte depletion module followed by a CD4+ T cell isolation channel greatly improves CD4 counting accuracy compared to the single channel device. Our technique establishes the ability to obtain CD4 counts directly from unprocessed whole blood with accuracy across the dynamic range of CD4 cell counts, and is suitable for simple, rapid and affordable HIV management in point-of-care and resource-limited settings.

580e "Detectorless" Electrophoresis for Multiplexed Enzyme Activity Assays

David Ross, National Institute of Standards and Technology, 100 Bureau Dr., MS 8313, Gaithersburg, MD 20899 and **Jason G. Kralj**, Biochemical Sciences, National Institute of Standards and Technology, 100 Bureau Dr., MS 8313, Gaithersburg, MD 20899.

A high-throughput separation system was developed by multiplexing gradient elution moving boundary electrophoresis (GEMBE), with each channel acting as separator and conductivity detector. GEMBE functions by combining electrophoresis with a pressure-driven bulk counterflow. The counterflow is gradually reduced until each analyte enters the separation channel and is detected. Each analyte appears as a step in the electrical current vs. time, which is smoothed and differentiated to produce electropherogram peaks. All measurements are electrical rather than optical, making the system easy to scale up to perform large numbers of simultaneous assays in parallel.

Using our prototype system, we demonstrated 16 parallel kinase enzyme activity assays with and without inhibitor; each sampled 13 times over 6 h to observe reaction progress. The other main advantages of this approach are that it excludes proteins and enzymes, eliminating the need for surface coatings or additives to reduce channel fouling; and no stop reagents are needed because the samples are analyzed in real time and parallel.

580f Self-Loading Lithographically Structured Microcontainers: 3D Patterned, Mobile Microwells

Timothy Leong[1], Bryan Benson[1], Christina Randall[2], Aasiyeh Zarafshar[1], and David H. Gracias[1]. (1) Department of Chemical and Biomolecular Engineering, Johns Hopkins University, 125 Maryland Hall, 3400 N Charles Street, Baltimore, MD 21218, (2) Department of Biomedical Engineering, Johns Hopkins University, Baltimore, MD 21218

We describe photolithographically structured, mobile microcontainers that load themselves as they self-assemble en masse at low temperatures (~40°C); these containers enable a method of parallel loading at biologically-compatible temperatures. The assembly of the microcontainers is thermally triggered, without the need for any external connections. This untethered nature, coupled with the use of ferromagnetic materials in the microcontainers, allows for remote-controlled guidance. Additionally, photolithography enables the containers to be fabricated with precisely-engineered monodisperse sizes, shapes, and surface porosity. We demonstrate the self-loading of glass beads, and to highlight the low temperature process, cells and embryos.

580g Electrochemical Detection of Signaling Biomolecules In Nanofluidic Devices

Edgar D. Goluch, Bernhard Wolfrum, Marcel A.G. Zevenbergen, and Serge G. Lemay. Kavli Institute of Nanoscience, Section Molecular Biophysics, Delft University of Technology, Lorentzweg 1, 2628 CJ Delft, Netherlands

The sensitive and selective detection of signaling biomolecules is an important component of medical research, with applications ranging from developmental biology to degenerative diseases. Several neurotransmitters, such as serotonin, dopamine, and norephinephrine, are redox active catecholamines that can be electrically detected. However, electrochemical detection of biological samples is limited since interfering species, such as ascorbic acid, mask the signal of the target molecules. We have developed a nanofluidic device that amplifies and selectively detects the electrochemical signal of catechols. The detection is based on rapid redox-cycling between electrodes inside of a nanochannel. This device is capable of detecting the signal generated by only a few hundred molecules residing in the active area of the nanofluidic sensor. The nanochannel design of the sensor eliminates interference by molecules undergoing irreversible redox reactions. We demonstrate the selectivity of the device by detecting catechol in the presence of ascorbic acid, whose oxidized form is only stable on the order of milliseconds.

Detection of Kinase Translocation Using Microfluidic Electroporative Flow Cytometry

Jun Wang, Ning Bao, Leela L. Paris, Hsiang-Yu Wang, Robert L. Geahlen and Chang Lu[a)]
Departments of Agricultural and Biological Engineering, Department of Medicinal Chemistry and Molecular Pharmacology, School of Chemical Engineering, Birck Nanotechnology Center and Bindley Bioscience Center, Purdue University, West Lafayette, Indiana 47907, USA.

I. INTRODUCTION

Translocation of a protein between different subcellular compartments is a common event during signal transduction in living cells. Integrated signaling cascades often lead to the relocalization of protein constituents such as translocations between the cytoplasm and the plasma membrane or nucleus. Such events can be essential for the activation/deactivation and biological function of the protein. The protein-tyrosine kinase, Syk, is a prime example of a protein that translocates to the B cell receptors (BCR) in plasma membrane as part of its role in signal transduction upon stimulation.

Determination of the translocation of kinases within cells has been traditionally carried out using methods such as subcellular fractionation/Western blotting or imaging of a few cells where only average information was gathered. Flow cytometry has been the tool of choice for single cell studies within cell populations of relatively large sizes. However, conventional flow cytometry is intrinsically insensitive to the subcellular location of the probed protein. Laser scanning cytometry (LSC) quantifies nuclear/cytoplasmic distribution of a fluorescent labeled protein based on solid phase cytometry technique [1-3] at a throughput only at ~100 cells/s. A recently developed complementation assay for protein translocation (CAPT) demands complex biological molecule manipulation [4]. Simple and robust high-throughput methods for examining intracellular translocation at the single cell level are highly demanded for mechanistic studies and clinical applications.

Here we demonstrate a new high-throughput technique, which we refer to as electroporative flow cytometry (EFC), to observe the translocation of the kinase Syk from the cytoplasm to plasma membrane at the level of the cell population with information gathered from single cells. Electroporation occurs when cells experience an electrical field with the intensity beyond a certain threshold. During electroporation the electrical field opens up pores in the cell membrane. Such pores allow the release of intracellular materials into the surrounding solution [5-7]. In this study, we found that the amount of SykEGFP left in the cells after electroporation was related to whether or not translocation of Syk to the plasma membrane occurred. Microfluidic EFC was able to detect the translocation of Syk and provide characteristics of the entire cell population in terms of the release of the intracellular kinase. We envision that the same mechanism can be extended to the detection of translocations involving other kinases and cell types.

II. EXPERIMENT

1. Microchip fabrication.

The general information about PDMS microfluidic chip fabrication based on standard soft lithography was provided in our previous publication[8]. The dimensions of fabricated microfluidic channels is showed in Figure 1 (a). The depth of the microfluidic device was uniformly 33 μm.

2. Cell sample preparation.

Syk- and Lyn-deficient (Syk/Lyn-deficient)[9] chicken DT40 B cells (DT40-Syk^--Lyn^-) and SykEGFP-DT40-Syk^--Lyn^- cell lines were produced as described before [10]. Cells can be stimulated with 50 μg/ml goat anti-chicken immunoglobulin M (IgM) antibody (Bethyl Laboratories, Montgomery, TX) for 1 h at 20 ℃.

To establish a negative control, both stimulated and unstimulated DT40-Syk^--Lyn^- were labeled with a fluorogenic dye, calcein AM (Invitrogen, Carlsbad, CA) after the above procedure for 1 h stimulation or incubation. The labeling was done by incubating the cells with calcein AM at a concentration of 20 ng/ml for 10 min.

3. Microchip operation and signal detection.

The setup of the apparatus is shown in Figure 1. The microfluidic EFC device was mounted on an inverted fluorescence microscope (IX-71, Olympus, Melville, NY) with a 40X dry objective (NA=0.60). A high voltage power supply (PS350, Stanford Research Systems, Sunnyvale, CA) was used to generate a constant direct current (DC) voltage in between the sample and the outlet reservoirs.

An air-cooled 100mW argon ion laser (Spectra-Physics, Mountain View, CA) at 488 nm was applied as the excitation source for laser-induced fluorescence that was recorded by LabView software (National Instruments, Austin, TX). The data were processed by programs written in MATLAB to extract histograms of the fluorescence from a cell population. The data were presented in 4 decades (from 0.001 to 10 V) logarithmic histograms with 256 channels. The voltage signal ranging from 1 mV to 10 V was converted to 4 decade logarithmic voltage scale and then 256 scale channels.

4. Conventional flow cytometry.

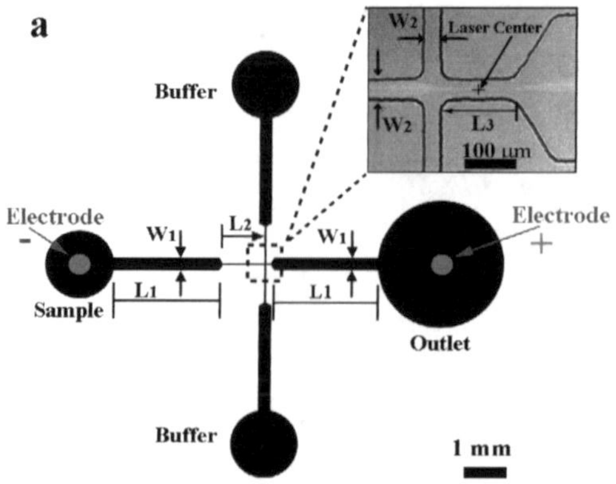

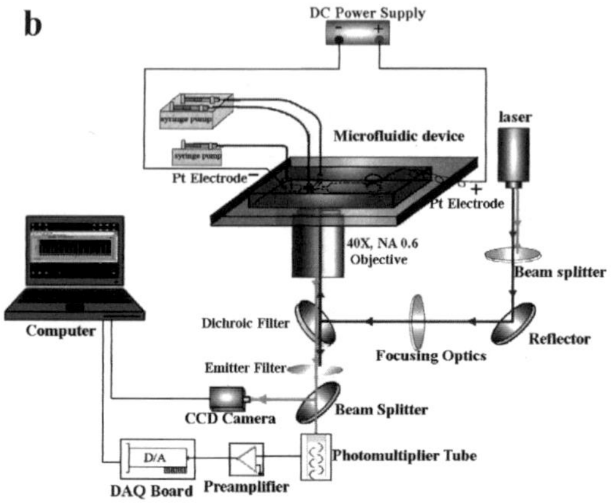

Figure 1. (**a**) The layout of the microfluidic chip. The width varied in the horizontal channel with W_1 of 300 μm and W_2 of 33 μm. The length of the narrow section L_2 was set as either 1 or 2 mm in different experiments. The other sections in the horizontal channel had a length of 2.5 mm for L_1 and 150 μm for L_3. The depth of the microfluidic channels was uniformly 33 μm. The laser detection point was positioned at the center of the horizontal channel after hydrodynamic focusing. In the inset image, the fluorescent trail was left by SykEGFP-DT40-Syk^--Lyn^- cells when the ratio between the flow rate in one of the two vertical channels and that in the horizontal channel was 3:1. (**b**) A schematic illustration of the setup of the microfluidic electroporative flow cytometry apparatus.

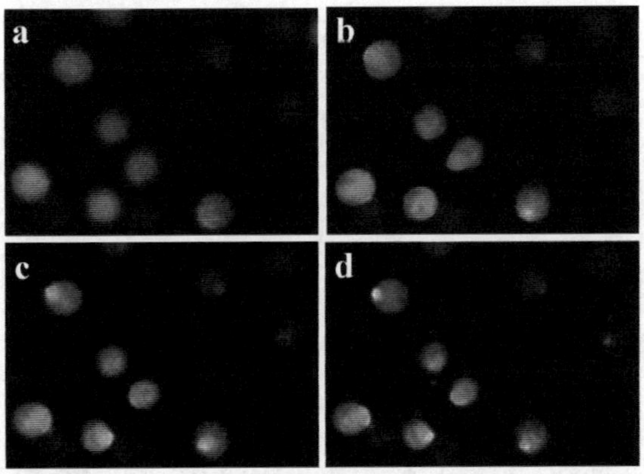

Figure 2. The fluorescent images showing capping of SykEGFP at the plasma membrane. SykEGFP was redistributed to the plasma membrane after stimulation with anti-IgM antibody at room temperature for 5 min (**a**), 30 min (**b**), 60 min (**c**) and 90 min (**d**). The cells labeled by red asterisks in (d) were considered the ones with capping. The percentage of cells with capping was 81±3% (enumerated in three separate experiments by one observer with the results of 396/470, 462/575, and 388/497).

We also conducted screening of the chicken DT40 cells (stimulated and unstimulated) using a commercial flow cytometer with 488 nm excitation and FL-1 525 nm BP emission filter (Cytomics™ FC 500 flow cytometer, Beckman Coulter, Inc., Fullerton, CA) at the Purdue University Flow Cytometry Laboratories. For each run, around 10000 cells were screened for fluorescence intensity.

5. Fluorescence intensity calibration.

We used Molecules of Equivalent Soluble Fluorophores (MESF) units to quantify the fluorescence intensity from single cells when microfluidic EFC was used [11, 12]. The mean percentage of EGFP or calcein of a cell population, P_{EGFP}, (by assuming that SykEGFP-DT40-Syk$^-$-Lyn$^-$ population without electroporation has a mean percentage of 1 and DT40-Syk$^-$-Lyn$^-$ population has a mean percentage of 0) was calculated using following formulae (Φ=fluorescence quantum yield, ε=extinction coefficient, A=absorbance, peak A_{EGFP}=489 nm, peak $A_{Fluorescein}$=495 nm, V=average cell volume, $EGFP_{mean}$= the population mean MESF value of EGFP fluorescence signal interpolated from the standard curve (Figure 5 in Supporting Information),

$$[EGFP] = EGFP_{mean} \times B \times K/V \quad (1)$$

Where $B = \Phi_{EGFP} \times \varepsilon_{EGFP} \times A_{EGFP:488nm}/A_{EGFP:484nm}/(\Phi_{Fluorescein} \times \varepsilon_{Fluorescein} \times A_{Fluorescein:488nm}/A_{Fluorescein:490nm})$

From the standard curve, K= moles fluorescein/relative fluorescent units. $[EGFP]_{control}$ is the value for DT40-Syk$^-$-Lyn$^-$ population (without EGFP tagging and without electrical field), $[EGFP]_{E=0}$ is the value for SykEGFP-DT40-Syk$^-$-Lyn$^-$ population without the electrical field. By assuming constant B, K and V under different electric field, P_{EGFP} of a cell population can be calculated to be

$$P_{EGFP} = \frac{([EGFP]-[EGFP]_{Control})}{([EGFP]_{E=0}-[EGFP]_{Control})} \times 100\% \quad (2)$$

The average percentage of calcein in a given cell population was calculated using the same method.

III. RESULTS AND DISCUSION

The design and setup of the microfluidic EFC device is shown in Figure 1. A new microfluidic electroporation technique based on constant DC voltage was applied here [7, 13]. The horizontal channel was composed of two wide sections (W_1 = 300 μm) and one narrow section (W_2= 30 μm) with the depth of the whole channel being uniform (Figure 1a). The field strength in the narrow section was approximated to be 10 times

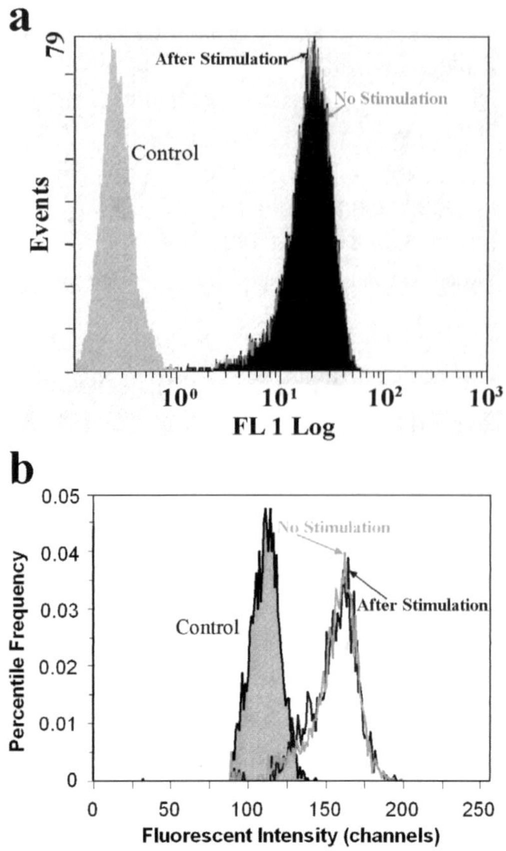

Figure 3. The histograms of the fluorescent intensity of SykEGFP-DT40-Syk⁻-Lyn⁻ cells with and without stimulation by anti-IgM antibody and the DT40-Syk⁻-Lyn⁻ cells (the control). (**a**) Histograms obtained by Cytomics FC 500 Flow cytometer (grey: without stimulation; black: stimulated by anti-IgM; Grey: the control, DT40-Syk⁻-Lyn⁻ cells without EGFP labeling). The data generated by dead cells (determined based on scatter properties) were excluded from the analysis. (**b**) Analysis of the same samples as in (**a**) in the microfluidic device (with the voltage at 0). To eliminate the effect of different cell sample sizes on the shape of the histograms, the distributions were normalized to have percentile frequency for the y-axis.

higher than the field strength in the wide sections according to Ohm's law, We have shown that when the overall voltage is in the right range, electroporation exclusively occurs in the narrow section since the field intensity in the wide sections are too weak to compromise the cell membrane [7, 13]. The velocity of cells here was determined mostly by the infusion rate of the syringe pump.

In our experiment, we applied SykEGFP-expressing chicken DT40 cells lacking both Syk and Lyn to ensure that the localization of Syk at the plasma membrane lasted long enough for us to finish the tests. The cells were stimulated at room temperature (20° C) by anti-IgM antibody. Fluorescence images were taken at timed intervals of 0.5 h. Figure 2 shows that the cells started to show patches and caps of fluorescent clusters at 5 min (Figure 2a). Within 60 min the vast majority of cells had a single cap at one pole. Figure 3 shows the event frequency versus the fluorescence intensity for the cell populations with and without anti-IgM stimulation. With a commercial flow cytometer the mean fluorescence intensity of the cell population with translocation (after anti-IgM stimulation) was 19.6 compared to 19.2 for the cell population without stimulation, on a 0.1 to 1000 logarithmic relative brightness scale (10,000 cells in each population). The histograms of the two cell populations overlapped very well. When screening the cells in our microfluidic device, the two cell populations (2,000~3,000 cells in each population) with and without stimulation also overlapped with the mean channels being 159 and 160, respectively. The results indicate that the translocation cannot be detected with conventional flow cytometry approach using either a commercial flow cytometer or our microfluidic flow cytometer.

We applied microfluidic EFC to analyze the cell populations with and without anti-IgM stimulation, under varying voltages across the sample and outlet reservoirs. Figure 4 shows the mean fluorescence intensity of the cell population plotted against the field intensity for all the three experiments. It was found that the optimal field intensities for detecting translocation to the plasma membrane in a cell population was around 700 and 800 V/cm with a duration of 120 ms or 1000 V/cm with a duration of 60 ms (significantly different at P<0.01).

We have also taken both phase contrast and fluorescent images of the cells after being processed in the microfluidic EFC device. The cell size after the tests became increasingly larger when the field intensity increased, due to the influx of the surrounding solution during electroporation. The residual fluorescence from the cells decreased with increasing field intensity. No significant fragmentation of the plasma membrane was observed. There was some sign of possible membrane fragments only at the highest field intensity

IV. CONCLUSION

By combining electroporation with flow cytometry, we demonstrate microfluidic EFC as a new tool to differentiate cell populations with different activation states. Microfluidic EFC offers a simple and robust physical tool for detecting kinase translocation within the scope of the entire cell population. In principle, there are no intrinsic limitations that prevent our technology from having the throughput and the stability similar to those of

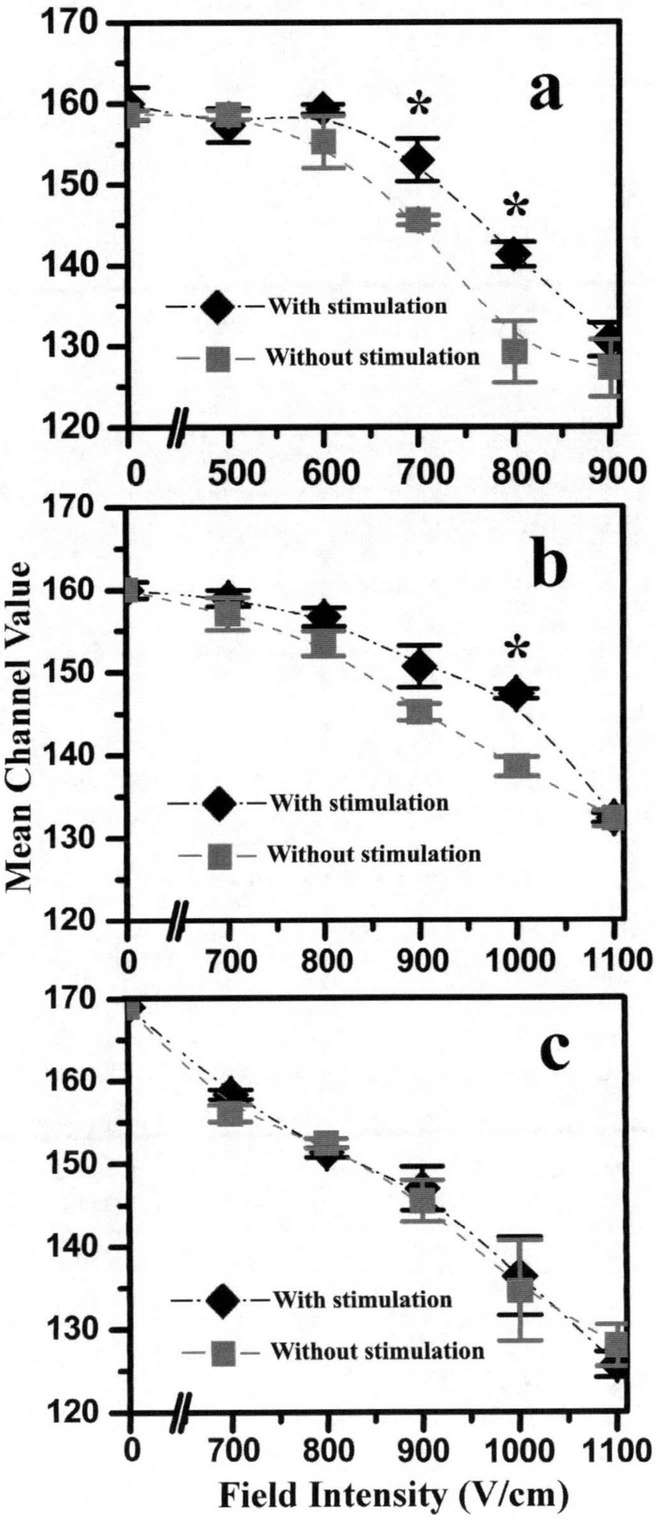

Figure 4. The variation of the mean fluorescence intensity value of the cell population at different field intensities with and without stimulation by anti-IgM. SykEGFP-DT40-Syk⁻-Lyn⁻ cells were applied in (**a**) and (**b**), while calcein AM stained DT40-Syk⁻-Lyn⁻ cells were used in (**c**). The data in (**a**) and (**b**) were obtained with different electroporation durations of 120 and 60 ms, respectively. The duration in (**c**) was 60 ms. The error bars were generated by carrying out the experiments in triplicate. (*) indicates significant difference at P<0.01, calculated using unpaired t test with equal variance.

conventional flow cytometry. The time required for the electroporative release of the intracellular materials from cells does not necessarily put a limitation on the throughput. In principle, the length of the narrow section can be increased to maintain enough duration for the electroporative release of intracellular materials while both the throughput and the velocity of the cells are high. The microfluidic platform is well suited for both research and clinical settings.

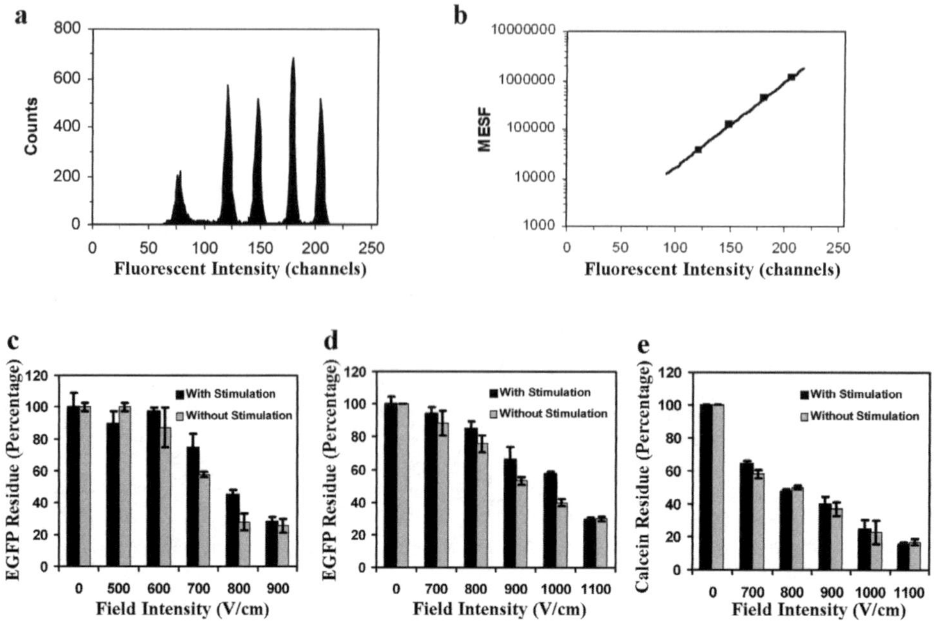

Figure 5. The quantitative analysis of the fluorescence intensity of DT40 cells. (**a**) The histogram of the fluorescence intensity from a mixture of beads with varying predefined fluorescence intensity levels. The data were generated by the microfluidic EFC system and the population with the lowest intensity is the reference blank. (**b**) The calibration curve with the MESF values of the beads plotted against the peak channel numbers (the fluorescent intensities) of the beads obtained by our microfluidic system (R^2=0.9978, solid square). The mean MESF value of the cells, $EGFP_{mean}$ in equation (5), is determined by finding the corresponding MESF value for a mean channel value known from the experiments. We assumed that the quantity of the fluorescent protein/calcein of the cell population detected by the system without the field (E = 0) was 100%. (**c**) The relative EGFP content in SykEGFP-DT40-Syk^--Lyn^- cells under different electric field intensities with a electroporation duration of 120 ms. (**d**) The relative EGFP content in SykEGFP-DT40-Syk^--Lyn^- cells under different electric field intensities with a electroporation duration of 60 ms. (**e**) The relative calcein content in calcein AM stained DT40-Syk^--Lyn^- under different electric field intensities with a electroporation duration of 60 ms.

V. ACKNOWLEDGEMENTS

We thank Purdue University and grant CA37372 from the National Cancer Institute for the financial support.

REFERENCES

1 Deptala, A., Bedner, E., Gorczyca, W. and Darzynkiewicz, Z. (1998) Activation of nuclear factor kappa B (NF-kappaB) assayed by laser scanning cytometry (LSC). *Cytometry* **33**, 376-382

2 Bedner, E., Li, X., Kunicki, J. and Darzynkiewicz, Z. (2000) Translocation of Bax to mitochondria during apoptosis measured by laser scanning cytometry. *Cytometry* **41**, 83-88

3 Ozawa, K., Hudson, C. C., Wille, K. R., Karaki, S. and Oakley, R. H. (2005) Development and validation of algorithms for measuring G-protein coupled receptor activation in cells using the LSC-based imaging cytometer platform. *Cytometry Part A* **65A**, 69-76

4 Wehrman, T. S., Casipit, C. L., Gewertz, N. M. and Blau, H. M. (2005) Enzymatic detection of protein translocation. *Nature Methods* **2**, 521-527

5 Rols, M. P. and Teissie, J. (1990) Electropermeabilization of mammalian cells. Quantitative analysis of the phenomenon. *Biophysical Journal* **58**, 1089-1098

6 Wang, H. Y. and Lu, C. (2006) Microfluidic chemical cytometry based on modulation of local field strength. *Chemical Communications*, 3528-3530

7 Wang, H. Y. and Lu, C. (2006) High-throughput and real-time study of single cell electroporation using microfluidics: effects of medium osmolarity. *Biotechnology and Bioengineering* **95**, 1116-1125

8 Wang, J. and Lu, C. (2006) Microfluidic cell fusion under continuous direct current voltage. *Applied Physics Letters* **89**, 234102

9 Takata, M. and Kurosaki, T. (1996) A role for Bruton's tyrosine kinase in B cell antigen receptor-mediated activation of phospholipase C-gamma 2. *Journal of Experimental Medicine* **184**, 31-40

10 Ma, H., Yankee, T. M., Hu, J. J., Asai, D. J., Harrison, M. L. and Geahlen, R. L. (2001) Visualization of Syk-antigen receptor interactions using green fluorescent protein: Differential roles for Syk and Lyn in the regulation of receptor capping and internalization. *Journal of Immunology* **166**, 1507-1516

11 Wang, L. L., Gaigalas, A. K., Abbasi, F., Marti, G. E., Vogt, R. F. and Schwartz, A. (2002) Quantitating fluorescence intensity from fluorophores: Practical use of MESF values. *Journal of Research of the National Institute of Standards and Technology* **107**, 339-353

12 Martin, B. R., Giepmans, B. N. G., Adams, S. R. and Tsien, R. Y. (2005) Mammalian cell-based optimization of the biarsenical-binding tetracysteine motif for improved fluorescence and affinity. *Nature Biotechnology* **23**, 1308-1314

13 Wang, H. Y. and Lu, C. (2006) Electroporation of mammalian cells in a microfluidic channel with geometric variation. Analytical Chemistry **78**, 5158-5164

14 Wang, J., Lu, C. (2006) Microfluidic cell fusion under continuous direct current voltage. *Applied Physics Letters* **89**, 234102

628b Droplet-Based PCR In a Valveless Microfluidic Device

Fang Wang and Mark A. Burns. Chemical Engineering, University of Michigan, 2300 Hayward, 3074 H.H.Dow, Ann Arbor, MI 48109

Polymerase chain reaction (PCR) is one of the most versatile molecular biological techniques, playing an indispensable role in many genetic analyses. Relying on microfabrication and microfluidic technologies, miniaturized PCR devices have been developed to achieve rapid amplification with reduced cost of reagents and samples. Certain issues such as evaporative loss and reagent adsorption, which stem from the characteristics of the thermocycled reaction and the micro format itself, have to be considered in designing and constructing the microdevices. Here we present a valveless microfluidic droplet-based PCR device that has no evaporative losses and has amplification efficiencies comparable to benchtop reactions. In this device, aqueous droplets containing the reaction solutions are encapsulated in a continuous oil phase. Single reaction droplets with a maximum volume of 200nL can be formed reproducibly within tens of seconds. With on-chip heaters and temperature sensors, PCR of λDNA has been successfully performed in less than half the time required for PCR in a thermocylcer. By using a FRET probe, the amplification can be detected in situ and in real time using fluorescent microscopy. The fluorescent intensity of the amplified products increases with the increase of template DNA concentration, and increases to a maximum value with the increase of cycle numbers. A higher polymerase concentration is still required in the droplet-based reaction to give on-chip amplifications comparable to benchtop results but the optimum Magnesium ion concentration is the same for both systems.

628c Parallel Multi-Time Point Cell Stimulus and Lysis In a Microfluidic Device Using Chaotic Mixing and Pressure Resistance

Alison Hirsch[1], Catherine Rivet[2], Boyang Zhang[1], Melissa Kemp[3], and **Hang Lu**[1]. (1) School of Chemical and Biomolecular Engineering, Georgia Institute of Technology, 311 Ferst Dr. NW, Atlanta, GA 30332, (2) Interdisciplinary Bioengineering Program, Georgia Institute of Technology, 311 Ferst Dr. NW, Atlanta, GA 30332, (3) The Wallace H. Coulter Department of Biomedical Engineering, Georgia Institute of Technology, 311 Ferst Dr. NW, Atlanta, GA 30332

To build a dynamics model of a complex cellular pathway, it is necessary to acquire a large data set, such as gene expression and protein activity at different time points upon stimulation by external signals. Many important protein activation events occur within minutes after stimulation. However, with conventional multi-well plate assays by manual handling, it is difficult to achieve adequate resolution in the appropriate time scales. Microfluidics is a capable alternative that can provide uniformity in sample handling to reduce error between experiments. We present a device for multi time-point lymphocyte stimulation and lysis for downstream analysis of protein activation. Mixing and even splitting of reagents into each time-point channel while using minimal amounts of reagents are key device features. Our device is capable of eight time points with controlled rapid mixing, precise timed stimulation, and rapid lysis. During operation, a syringe pump drives the flow to only three inlets (cells, stimulus, and lysis buffer), cells and stimulus are mixed and split into eight equal streams in the stimulation chip. The majority of the incubation time occurs in the tubing leading to the lysis chip, where the reaction is quenched and cells are ruptured with lysis buffer to extract intracellular proteins. Rapid mixing is essential to our design for precise definition of stimulation time. A staggered herringbone array was used to achieve full mixing of reagents with minimal shear in less than 0.2 sec on the stimulation chip (and <0.9 sec on the lysis chip), an extremely small fraction of the stimulation time. Because our reagents have different fluidic properties, a COMSOL model was created to probe the mixing effects of viscosity and density. Our model suggests the effect is solely due to viscosity, not the change in density. Further confocal microscopy experiments were used to visualize the mixing of solutions with mismatched viscosity. Equal flow rates in each stream were achieved by balancing the channel resistance: all channels have the same small dimensions with pressure drops orders of magnitude greater than the tubing. Although to set up multiple stimulation times we use different lengths of tubings (with varying inner diameters), the flow rates remain uniform. This configuration gives large flexibility of the specific time points we want to achieve. With commercially available tubings, we can easily achieve stimulation times ranging from 30 seconds to 1 hour. We show that multiple protein activation events in lymphocyte stimulation by antibodies can be detected in a much smaller number of cells using our system than in conventional assays.

Enhancement of acoustic streaming induced flow on a focused SAW device: Implications for biosensing and microfluidics

Reetu Singh, Subramanian K.R.S. Sankaranarayanan, and Venkat R. Bhethanabotla *
Sensors Research Laboratory, Department of Chemical and Biomedical Engineering, University of South Florida, Tampa, Florida, 33620, USA.

Abstract

Surface acoustic wave (SAW) devices used for biological species detection suffer from fouling that result from binding of non-specific protein molecules to the device surface. Non-specific binding dramatically reduces the sensitivity and selectivity of biosensors. Fluid motion induced from high intensity sound waves is called acoustic streaming. The acoustic streaming phenomenon can be used to remove these non-specifically bound proteins to allow reuse of SAW devices. Focused interdigital transducers (FIDTs) based on concentric wave surfaces can excite surface acoustic waves (SAW) with high intensity, high beam-width compression ratio and small localized area. The excited waves can be utilized to enhance the streaming induced removal of fouling proteins leading to highly sensitive biosensors.

In the present work, we have developed a 3-D FE-FSI model to investigate and analyze the streaming velocity fields and forces induced by SAW devices with focused Interdigital transducers based on concentric wave surfaces. The acoustic streaming fields and velocities are computed and analyzed. The simulated device surface displacements indicate an increased surface displacement of a focused SAW device compared to a conventional SAW device with a similar size, finger periodicity and applied input voltage. A comparison of the velocity fields between focused and conventional SAW devices indicates a significant increase in the tangential, normal, and consequently total streaming velocity as a result of focusing brought about by the FSAW devices, thereby making them more suitable for biofouling removal in biosensors compared to conventional devices.

I. Introduction

Surface acoustic wave (SAW) sensors find intensive applications in chemical and biological sensing owing to their portability, cost effectiveness, high sensitivity, and reliability [1-8]. Typically, biosensing applications require the detection and measurement of biomarkers in fluid media [9]; an example being the measurement of certain proteins in bodily fluids for the detection of ovarian cancer. The use of SAW sensors in biosensing applications requires the integration of SAW devices with microfluidics, which is the science of designing systems and processes that handle and use fluid volumes of the order of picoliters to nanoliters by reducing the dead volume in the system. In microfluidic applications, SAW devices can be used as actuators for pumping small volumes of fluid to generate high enough velocities in the small fluidic channels owing to the high intensity acoustic waves. Some applications of microfluidics include gene expression analysis using DNA chips, DNA hybridization, lab-on-chip systems, immuno-assays, micro-arrays, biosensing, drug screening, drug delivery, ultrasonic mixing, actuation, and flow cytometry to name a few [10-17]. Most of these clinical and diagnostic applications of microfluidic devices require transduction of picogram to nanogram level of biomarkers into a readable signal without interference from other proteins and biomarkers, thereby requiring high device sensitivity and selectivity. However, most of the biosensors are plagued with the issue of non-specific protein binding and therefore analyte discrimination

* Author for correspondence at email: venkat@eng.usf.edu

[18]. Thus, non-specifically bound (NSB) protein removal is a challenge in biosensing applications. Additionally, high device sensitivity is required so as to detect small enough concentrations of the analyte.

The piezoelectric crystal orientation, the Interdigital transducer (IDT) geometry/configuration, and crystal thickness determine the mode of the propagating waves. Rayleigh waves couple strongly with the fluid in contact with the sensor, launching high intensity acoustic waves into the fluid. The SAW-fluid interaction creates a pressure gradient in the direction of acoustic wave propagation in the fluid, leading to an acoustically driven streaming phenomenon known as SAW streaming [19-21] which can be used for removal of NSB proteins in order to improve sensor selectivity and sensitivity, as has been shown experimentally [22, 23]. The efficient utilization of streaming in SAW devices requires the understanding of fluid dynamics in these systems. Transducer designs that can bring about enhancement in acoustic streaming need to be identified. One such candidate is a focused SAW device, shown n Fig. 1, which can excite waves with high intensity, high beam-width compression ratio, small localized area, and high vertical displacement component [24]. This aspect of the focused SAW device can be used to increase streaming velocities and facilitate microtransport.

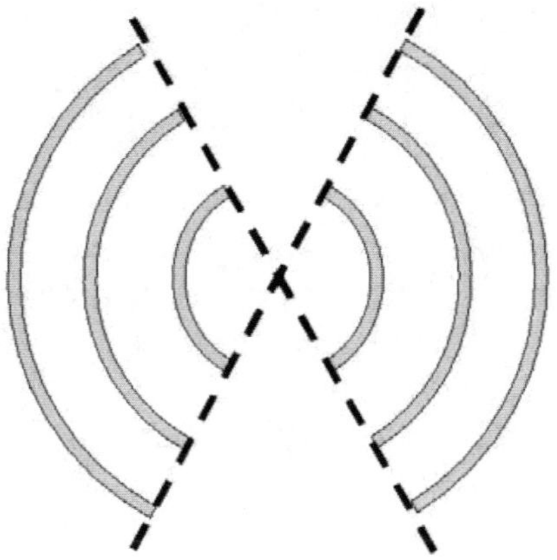

Figure 1: Schematic showing focused transducer used to enhance acoustic streaming

Structural simulations for a focused SAW device developed previously indicate increased amplitudes of surface acoustic waves in the area near the focal point, thereby making them suitable candidates for increasing the streaming capability [25]. Such models have been used in conjunction with parameters derived from the perturbation theory to predict streaming velocities and forces based on the continuum model of Nyborg [25]. However, these models ignore the effect of liquid loading and viscous dissipation in the presence of the same. The primary reason, being the simplification of the Navier-Stokes equation to ignore the viscous effects as adopted by most of the existing theories on acoustic streaming. To the best of our knowledge, there has been no three dimensional fluid structure interaction study on streaming induced by SAW devices which can quantitatively give an estimate of the streaming velocity variations introduced by viscous effects.

Understanding the interaction of the fluid field with the SAW sensor and the effects of various IDT configurations on wave propagation in the crystal requires the development of coupled field structural and FSI FE models which would enable investigation of methods to increase the acoustic streaming velocity for NSB removal while minimizing the influence of the streaming force on the sensing layer, thereby increasing the sensitivity and selectivity. The present study represents one of the first attempts to include the effects of liquid viscosity on the streaming velocity profiles. In the present work, we have developed a 3-D FE-FSI model to investigate and analyze the streaming velocity fields and forces induced by SAW device with FIDTs based on concentric wave surfaces. The acoustic streaming enhancement brought about by the focused SAW device is analyzed by comparison with a conventional SAW device having uniform IDTs.

II. Theory

A coupled-field FSI model of a SAW device based on a micron-sized piezoelectric substrate (YZ-$LiNbO_3$) in contact with a liquid loading was developed to study surface-acoustic-wave interaction with fluid loading. A system of four coupled wave equations for the electric potential and the three component of displacement in piezoelectric materials are solved for the piezoelectric substrate or the solid domain [26]:

$$-\rho\frac{\partial^2 u_i}{\partial t^2}+c_{ijkl}^{E}\frac{\partial^2 u_k}{\partial x_j \partial x_l}+e_{kij}\frac{\partial^2 \phi}{\partial x_k \partial x_j}=0 \tag{1}$$

$$e_{ikl}\frac{\partial^2 u_k}{\partial x_i \partial x_l}-\varepsilon_{ik}^{s}\frac{\partial^2 \phi}{\partial x_i \partial x_k}=0 \tag{2}$$

These coupled wave equations can be discretized and solved for generating displacement profiles and voltages at each element/node. The piezoelectric material displacements obtained from the above equations are applied to the fluid domain at each time step.

Fluid domain was modeled using the Navier-Stokes equation; the arbitrary-Lagrangian-Eulerian approach [27] was employed to handle the mesh distortions arising from the motion of the solid substrate.

$$\rho\left(\frac{\partial v_f}{\partial t}\right)+v_f\cdot\nabla v_f+\nabla P-2\eta\nabla\cdot D=0 \tag{3}$$

$$\nabla\cdot v_f=0 \tag{4}$$

Here, v_f, P, ρ and η denote the fluid velocity, pressure, density, and viscosity, respectively. D is the rate of deformation tensor given by

$$D=\frac{1}{2}\left(\nabla v_f+(\nabla v_f)^t\right)$$

The fluid-solid coupling was established by maintaining stress and displacement continuity at the fluid-structure interface. The velocity v calculated from Eq. (3) and (4) contains harmonically varying terms and a "dc" term. The latter induces acoustic-streaming. When averaged over a relatively long time, the effect of the harmonically varying terms disappears and only the contributions from the dc part appear in the solution. The acoustic-streaming velocity ($\bar{v}_{a,i}$, i=x, y, and z) is therefore obtained by averaging v over a time period as follows:

$$\bar{v}_{a,i}=\frac{1}{T}\int_0^T v_i dt\text{ , i=}x, y\text{ and }z \tag{2.5}$$

where T is the time period of the wave propagation.

III. Computational details

A SAW device based on YZ-$LiNbO_3$ with a liquid loading was modeled to gain insights into the acoustic streaming phenomenon, using a three dimensional coupled field FSI model developed for the first time in this work. The focused SAW (F-SAW) device was constructed by adopting a pair of concentrically shaped Focused IDTs (FIDTs) (Fig 2). The dimensions of the piezoelectric substrate were 400μm width x 500μm propagation length x 200μm depth. Two IDT finger pairs in each port were defined at the surface of Y-cut, Z-propagating $LiNbO_3$ substrate. The fingers were defined with periodicity of 40 μm and the aperture width of the fingers varies depending on their radial distance from device center. The IDT fingers were modeled as mass-less conductors and represented by a set of nodes coupled by voltage degrees of freedom (DOF). The transmitting and receiving IDT's are spaced 90 microns or 2.25λ apart. The model was meshed with tetragonal solid elements with four degrees of freedom, three of them being the three translations and the fourth being the voltage. To capture the dynamics at the interface and optimize on the simulation time, the mesh was defined such that it had high density at the interface and coarser away from it. A total of 2, 218, 399 nodes and 2, 085, 877 elements were generated.

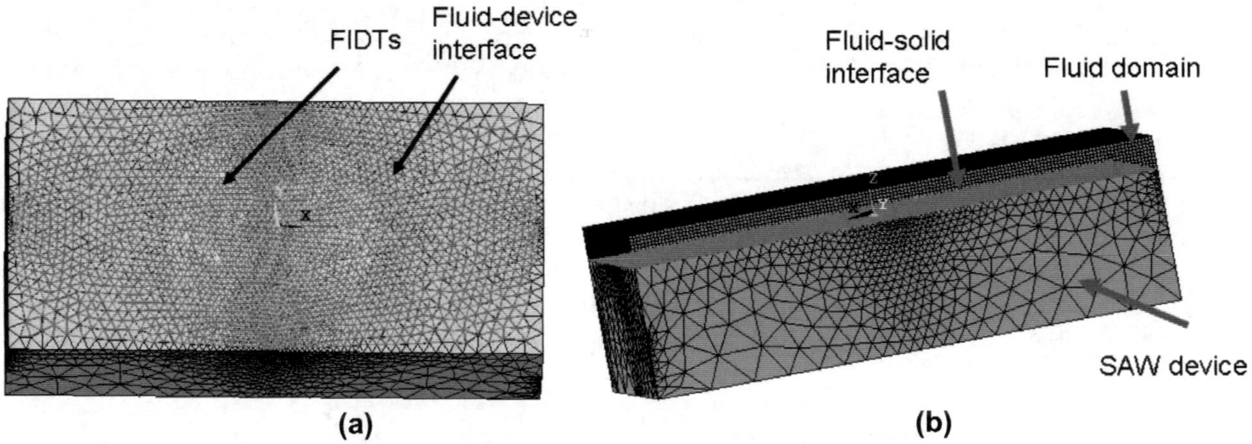

Figure 2. Focused surface acoustic wave device (a) meshed structure with FIDTs (b) with fluid loading

Fluid was modeled as incompressible, viscous, and Newtonian using the Navier-Stokes equation. In modeling fluid-solid interaction, fluid is described with reference to an Eulerian frame while the Lagrangian frame is more suited for structural/solid domain. However, the two frames are incompatible. This incompatibility is overcome by using the arbitrary Lagrangian-Eulerian (ALE) method where the mesh is constantly updated without modifying the mesh topology. In order to account for the fluid-solid interaction, an interface was defined across which displacements were transferred solid from to fluid and pressure from fluid to solid. The fluid mesh was be continuously updated as the piezoelectric substrate underwent deformation. The structure was simulated for a total of 100 nanoseconds (ns), with a time step of 1 ns. The excitation of the piezoelectric solid was provided by applying an AC voltage (with a peak value of 2.5 V and frequency of100 MHz) on the transmitter IDT fingers (Fig. 3).

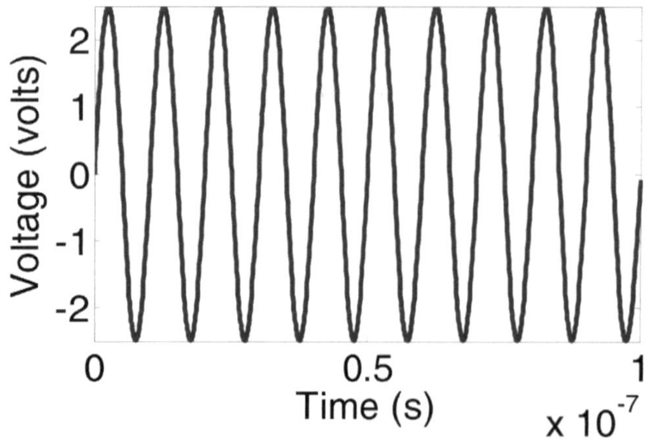

Figure 3: Applied voltage for structure excitation

IV. Results/Discussion

We have investigated wave propagation and acoustic streaming velocity fields for a focused transducer SAW device with a 120 degree arc and focal length of 85 μm, using the developed 3-D FSI model and compared them with a conventional SAW device having similar size, finger periodicity and applied input voltage to study the acoustic streaming enhancement brought about focused IDTs. Figures 4 (a, b) show the simulated device displacement contours at 70 ns, for the FSAW device as well as the conventional SAW device. Our results indicate that the maximum displacements in a FSAW device are greater than those in a conventional SAW device. In addition, the displacement field for a FSAW device tends to converge at the focal point with the maximum occurring at the focal point located at the center of the delay path and progressively decaying away from the focal point along the delay path. The acoustic energy in a FSAW device is mainly confined to the localized region near the center of the SAW delay path.

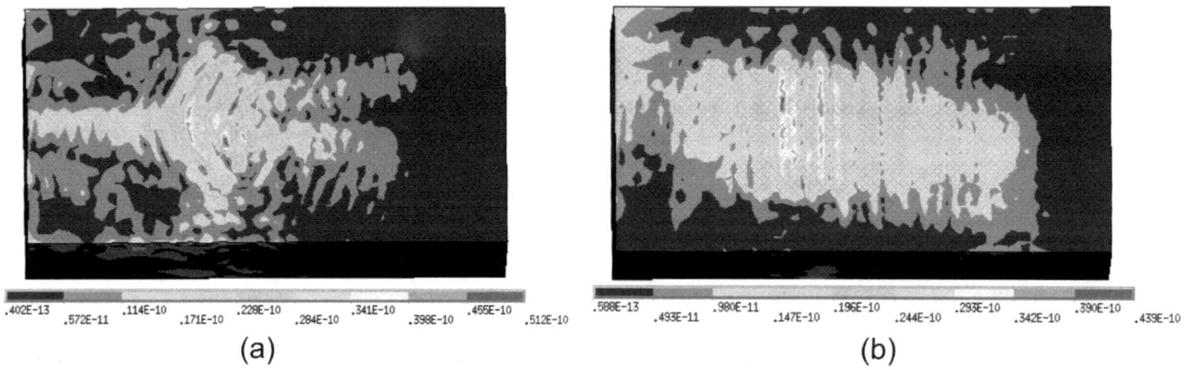

Figure 4: Contours showing the device surface displacement in response to an applied AC voltage at input IDTs at t= 70 ns for (a) FSAW device (b) conventional SAW device

Equation (5) was used to compute the streaming velocity in the F-SAW device. The calculated streaming velocity obtained for an F-SAW was compared to a conventional SAW device with the same wavelength. Figures 5(a, b) compare the tangential velocity profiles for a conventional and F-SAW device in the propagation and transverse directions; the normal and total streaming velocity profiles are compared in Figs. 6(a, b). The velocity profiles for both the devices indicate that the highest tangential and normal velocities occur close to the device surface. The fluid motion is confined to the initial few fluid layers beyond which the wave motion is dampened significantly. In addition, beyond the first few layers, flow reversal is observed indicating fluid recirculation close to the SAW device surface. Further, the results indicate an increase in fluid velocity in both the tangential and normal directions due to the focusing effect of a FSAW device as compared to a conventional SAW device. The FSAW device brought about a 32.3% increase in total streaming velocity compared to the conventional device; the FSAW induced increase in tangential velocity was 351.8% (in direction of wave propagation) and 216.1% (in transverse direction) and the normal velocity increased by 352.7%. These results suggest that FSAW device leads to an increase in the total acoustic streaming velocity thereby leading to an increase in NSB removal efficiency, thus making the FSAW device more suited to biosensing applications.

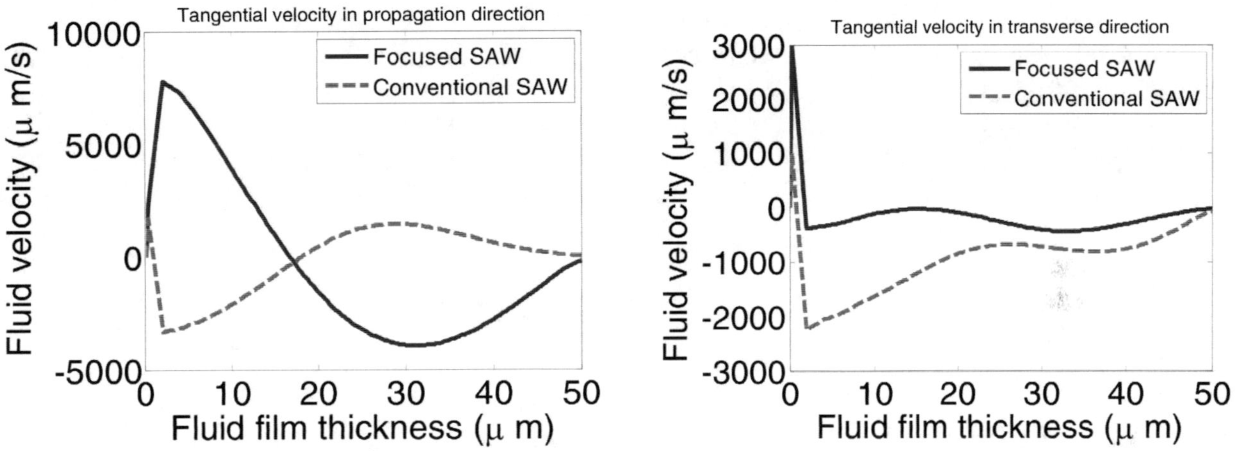

Figure 5: Comparison of tangential velocities (a) propagation direction (b) transverse direction for Focused SAW vs. Conventional SAW device for a peak input AC voltage of 2.5 V

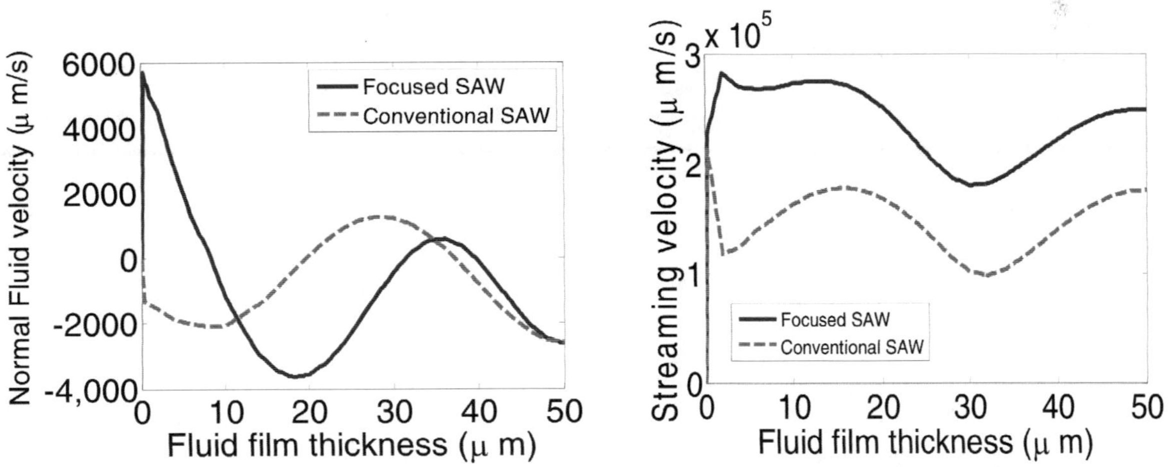

Figure 6: Comparison of (a) Normal fluid velocity (b) total streaming velocity for a Focused SAW vs. Conventional SAW device for a peak input AC voltage of 2.5 V

Conclusions

In this work, a 3D FE FSI model of an F-SAW device was developed to study the enhancement in acoustic streaming brought about by focusing. Comparison of velocity fields generated by a FSAW device, with the conventional SAW devices fabricated with uniform interdigital transducers suggests that the focused SAW devices leads to increased induced acoustic streaming velocity, thereby facilitating increased removal of non-specifically bound fouling proteins and making it more suited for biosensing applications. Further work is underway to compute and compare the streaming forces from a FSAW device with a conventional one to estimate the FSAW induced enhancement in biofouling removal.

Acknowledgements

The authors gratefully acknowledge the computational resources provided by Academic computing and Engineering computing at the University of South Florida, Tampa, FL. This work was supported by NSF grants CHE-0722887 and ECCS-0801929, and IIP-07122360.

References

1 E. Gizeli, C. R. Lowe, M. Liley, et al., Sensors and Actuators B 34, 295 (1996).
2 F. Josse, F. Bender, and R. W. Cernosek, Anal. Chem. 73, 5937 (2001).
3 Z. Li, Y. Jones, J. Hossenlopp, et al., Analytical Chemistry 77, 4595 (2005).
4 M. Hoummady, A. Campitelli, and W. Wlodarski, Smart Materials Structures 6, 647 (1997).
5 J. W. Grate, S. J. Martin, and R. M. White, Anal. Chem. 65 (1993).
6 J. W. Grate, S. J. Martin, and R. M. White, Anal. Chem. 65 (1993).
7 D. S. Ballantine and H. Wohltjen, Anal. Chem. 61, 704A (1989).
8 J. W. Grate, S. W. Wenzel, and R. M. White, Analytical Chemistry 63(15), 1552 (1991).
9 F. Josse, F. Bender, R. W. Cernosek, et al., IEEE International Frequency Control Symposium and PDA Exhibition, 454 – 461 (2001).
10 Z. Guttenberg, A. Rathgeber, S. Keller, et al., Physical Review E: Statistical, Nonlinear, and Soft Matter Physics 70, 056311/1 (2004).
11 A. G. Hadd, D. E. Raymond, J. W. Halliwell, et al., Analytical Chemistry 69, 3407 (1997).
12 A. Hatch, A. E. Kamholz, K. R. Hawkins, et al., Nature Biotechnology 19, 461 (2001).
13 K. Lange, G. Blaess, A. Voigt, et al., IEEE international ultrasonics, ferroelectrics, and frequency control joint 50th anniversary conference, 321 (2004).
14 K. Lange, F. Bender, A. Voigt, et al., Analytical Chemistry 75, 5561 (2003).
15 J. Bennès, S. Alzuaga, S. Ballandras, et al., IEEE Ultrasonics symposium, 823 (2005).
16 M. Alvarez, J. Friend, L. Yeo, et al., 16th Australian fluid mechanics conference, 621 (2007).
17 Z. H. Fan, S. Mangru, R. Granzow, et al., Analytical Chemistry 71 (1999).
18 C. M. Gregorya, J. V. Hatfielda, S. Higginsa, et al., Sensors and Actuators B: Chemical 65, 305 (2000).
19 L. Byoung-Gook, L. Dong-Ryul, and K. Kwon, Applied physics letters 89, 183505 (2006).
20 K. Chono, N. Shimuzu, Y. Matsui, et al., Japanese journal of applied physics 43, 2987 (2004).
21 K. Chono, N. Shimizu, Y. Matsui, et al., IEEE Ultrasonics Symposium 2, 1786 (2003).
22 S. Cular, D. W. Branch, V. R. Bhethanabotla, et al., AIChE Annual Meeting. (2005b).
23 S. Cular, D. W. Branch, V. R. Bhethanabotla, et al., IEEE Sensors Journal 8, 314 (2008b).
24 J. M. M. de Lima, F. Alsina, W. Seidel, et al., Journal of Applied Physics. 94, 7848 (2003).
25 S. K. R. S. Sankaranarayanan and V. R. Bhethanabolta, Journal of Applied Physics 103, 064518/1 (2008).
26 D. S. Ballantine, S. J. Martin, A. J. Ricco, et al., *Acoustic Wave sensors: Theory, design, and Physico-chemical applications* (Academic Press, San Diego, 1997)
27 J. Donea, S. Giuliani, and J. P. Halleux, Computer methods in applied mechanics. 33, 689 (1982).

628e Design and Evaluation of Polymeric Microfludic Biochips for Enzyme-Linked Immunosorbent Assay

Hongyan He[1], Yuan Yuan[2], Nan-rong Chiou[1], and L. James Lee[1]. (1) Department of Chemical and Biomolecular Engineering, The Ohio State University, 140 WEST 19TH AVE, Columbus, OH 43210, (2) Nanoscale Science and Engineering Center for Affordable Nanoengineering of Polymeric Biomedical Devices, The Ohio State University, 140 WEST 19TH AVE, Columbus, OH 43210

Enzyme-linked immunosorbent assays (ELISA), one of the most common immunoassays, is widely used for detection and quantification of chemical and biological molecules and is becoming more and more important in clinical diagnostics. To reduce the experimental time and reagent quantities, a polymeric microfluidic biochip has been designed and fabricated based on the integration of manufacturing modules: polyaniline-based surface modification, the conceptual design with splitters, and the oxygen plasma-PEI-TR protein A modification. The proper flow sequencing was achieved on a CD-like microfluidic chip by integrating the necessary microfluidic functions such as capillary valving treated with polyaniline-based surface modification. It has been successfully demonstrated that the splitters indeed work very well to even distribution of the fluid to branch flows. To enhance the sensitivity of miniature detection area on the PMMA biochips for ELISA immunoassay, a TR-catalyzed protein A antibody immobilization technique was successfully developed. By using this method, not only the antibody binding efficiency, but also the specific capture capacity towards target protein of the antibody was greatly enhanced, leading to a significant improvement of the fluorescence signal ELISA in PMMA chips including a lower detection limit, a wider linear detection range, a much shorter assay time comparable to 96-well microtiter plates. The microfluidic biochips we developed here would be applicable to a variety of clinically relevant disease conditions. And the modification technologies in this study can be extensively implemented in lab-on-a-chip systems, drug/gene delivery carriers and other immunoassay biosensor applications.

628f On-Chip Analysis of Neutrophil Isolation and Migration In Complex Chemotactic Environments

Nitin Agrawal, Daniel Irimia, and Mehmet Toner. Harvard Medical School, 114 16th st., Room 1239, Charlestown, MA 02129

Chemotaxis is one of the primary immunological responses against bacterial infection and plays a significant role in several physiological processes such as inflammation, wound healing, metastasis, atherosclerosis, arthritis etc. *In vivo*, the effective directed migration involves sequential events of intricate signaling and feedback mechanisms to guide the circulating neutrophils from blood stream to the site of injury. Recent studies suggest that neutrophil responses to chemoattractants following thermal injury are affected by a variety of cytokines/chemokines secreted by the injured tissue or the bacteria itself. Despite the fact that many of the involved signaling molecules have already been characterized, the global mechanisms through which neutrophils respond to heterogeneous microenvironment are not well understood. Several traditional approaches have been suggested to perform *in vitro* chemotaxis studies in artificially created chemotactic gradients. However, most of these approaches have limitations of gradient nonlinearity, spatial and temporal instability that are hard to replicate under similar conditions. Moreover, they involve extensive processes to isolate neutrophils (e.g. ficol separation) from large volumes of whole blood (of the order of milliliters) allowing their prolonged exposure to unfavorable conditions that could adversely affect their overall migration performances.

We hypothesize that the sepsis complications in burn patients occur due to the improper activation of neutrophils ultimately causing multiple organ failure. We study their migration behavior using microfluidic tools capable of performing fast switching between different chemokines in the gradient chamber (within ~ 5 seconds). An advanced switching gradient device incorporating microstructured valves has been developed that can be used to directly isolate primary neutrophils from just a drop (<10μL) of whole blood as well as provides a robust platform to perform chemotaxis assays in the competing environment of different chemokines. A drop of blood obtained from finger prick mixed with heparin solution is loaded in the flow chamber individually pre-coated with specific cell adhesion molecules (in this case P-selectin or E-selectin) that facilitate cell adhesion and capturing. We observe low affinity binding and rolling action of leukocytes in the absence of a stimulus while the rolling action turned into firm adhesion inside chemokine-pretreated channels. Observations from neutrophil capture characterization suggest that the P-selectin and E-selectin substrates demonstrate optimum capture efficiencies of 30-40 cells/mm^2 at concentrations 25μg/mL and 50μg/mL respectively. We also performed migration experiments on these substrates under individual gradients of chemoattractants fMLP and IL-8 and the observed average velocities indicate comparable migration patterns. Similar experiments will be performed (both pre-burn and post-burn) in an overlapping environment of different chemokines (IL8, C5a, GRO-α, fMLP and LTB4) and the migration behaviors will be compared. This study could open the possibility of using simplified and inexpensive point-of-care diagnostic methods based on microfluidic devices for predicting clinical outcome in burn patients.

Insulator-based Dielectrophoretic Lab-on-a-Chip System for Erythrocytes

Soumya S. Keshavamurthy, Mississippi State University, Mississippi State, MS, USA
Shirley Anell Pullen, Mississippi State University, Mississippi State, MS, USA
Adrienne R. Minerick, Mississippi State University, Mississippi State, MS, USA

Abstract

Medical lab work, such as blood testing, will soon be near instantaneous and inexpensive via capabilities enabled by the fast growing world of microtechnology. Previous research by the authors has shown that erythrocytes of different blood types respond differently in AC dielectrophoretic fields. These results suggest that spatial separation in the electric field depends on the blood type antigen expressed on the surface of the erythrocytes. This attribute forms the foundation for the continuous flow, DC dielectrophoretic research presented here. In this work, no electrodes are placed within the microchannel, thus simplifying fabrication and reducing electrode/cell interactions. Instead, a DC electric field is applied remotely and an insulating obstacle in the microfluidic channel creates a spatially non-uniform electric field. In our work, we utilize PDMS obstacles of rectangular, trapezium and semi-circle geometrical configurations on the order of 100 μm in breadth. Device dimensions were optimized by evaluating the behaviors of fluorescent polystyrene particles of three different sizes roughly corresponding to the three main components of blood: platelets (2-4 μm), erythrocytes (6-8 μm) and leukocytes (10-15 μm). This work provides the operating conditions for which size dependent blood cell DC dielectrophoresis can be performed. The optimized device is being used for continuous separation of erythrocytes. The design enables collection into specific channels based on the cell's deflection from the high field density region of the obstacle. By monitoring cell accumulation in the outlet ports, we expect to be able to infer blood type for unknown samples of blood. If successful, this developed technique can be directly applied for use in portable blood diagnostic devices for easy, accurate and rapid analysis.

Introduction

Lab-on-a-chip systems have the capacity to perform a variety of tasks ranging from DNA analysis to protein recognition and can also be catered to point-of-care medical diagnostic tools. Lab-on-a-chip (LOC) devices integrate multiple laboratory functions on a single square centimeter chip and are capable of handling extremely small fluid volumes down to picoliters [1-4]. They commonly utilize electrokinetics to move analytes because electric fields are versatile and can be precisely controlled for specific, quantifiable analyte responses. Furthermore, devices employing electric fields can eventually be simplified to only require a battery for power – a key characteristic for true portable diagnostic devices. One type of electrokinetics, dielectrophoresis (DEP), uses spatially non-uniform alternating current (AC) electric fields, which controls particles or cells through transient polarizations. Dielectrophoresis is a key type of field to explore in medical diagnostic devices due to operational simplicity, requirement of lower voltages, and small sample volumes all of which enable device portability [2].

Direct Current dielectrophoresis (DC-DEP) or insulator-based dielectrophoresis (iDEP) is a relatively new field. It utilizes the spatially nonuniform electric field component but not the frequency dependent component of dielectrophoresis, which has been shown to be adequate for manipulation of cells [5]. Spatial non-uniformities in the electric field are generated via insulating obstacles positioned in microdevice channels through which a DC current is passed. The advantages of using insulating obstacles include [6]:

1. Insulators are less prone to fouling than electrodes embedded in microdevices,

2. No metal components are involved which reduces the complexity of fabrication of devices,
3. Ideal insulating obstacles are mechanically robust and chemically inert, and
4. Electrolysis gas evolution at the remotely located metal electrodes does not result in bubbles inside the channel.

Initial work on iDEP was conducted by Cummings and Singh in 2003. Their configuration contained an array of insulating posts where two operating regimes were observed, namely streaming and trapping DEP [7]. In streaming DEP, particles concentrated at electric field maxima, travel down in flowing streams, and in trapping DEP the forces reversibly immobilize the particles on the insulating posts. Subsequently, Lapizco-Encinas *et al.* used insulating posts to concentrate and sort live and dead *E.coli* [5] where live cells exhibited trapping DEP and dead cells exhibited streaming DEP. They also demonstrated that dead cells exhibited less negative DEP compared to live cells. In these cases, size was maintained constant while cell properties varied. There were many efforts to explore other insulating materials for iDEP included cyclo-olefin polymers by Mela *et al.* in 2005 [8], and an oil droplet obstacle [9] in 2006. Thwar *et al.* (2007) demonstrated dielectrophoretic potential wells using pairs of insulating oil menisci to shape the DC electric field [10] and Zhang *et al.* simulated and modeled separation in a circular channel [11]. Fabrication of these microdevices have been achieved using silica or glass, thermal plastics such as polymethylmethacrylate (PMMA), polyolefins, polyethylene terephthalate, and polycarbonate or elastomers such as poly(dimethylsiloxane) (PDMS) [12]. A number of insulating obstacles have been explored by other researchers that comprise different shapes like rectangle [13], triangle [13], circular [12] and saw-tooth [14]. A complete picture of shape and voltage dependencies in those devices has not yet been attained.

This paper discusses a novel premise of using DC dielectrophoresis to ascertain cell behaviors for the blood types of the ABO system (A^+, B^+, AB^+, O^+, A^-, B^-, AB^-, and O^-). Previous research by the authors has shown that blood types respond differently in AC dielectrophoretic fields. Dielectrophoretic movement for the red blood cells followed distinct trends with time for each blood type and was quantified by these parameters: total cell count, vertical movement, horizontal movement, and distance. All parameters at 95% confidence interval revealed that type O^+ was easily distinguishable from the other blood types with types A^+ and B^+ having similar responses in the dielectrophoretic field. Blood type A^+ deflects from high field density region followed by type B^+ and AB^+, whereas type O^+ displays small deflections in the nonuniform 1 MHz electric field with 0.025 V_{pp}/μm strength [3]. These observed cell blood type deflections have been attributed to the functionalized ABO antigens on cell membranes [3]. These results suggest that spatial separation in the electric field depends on the blood type antigen expressed on the surface of the erythrocytes.

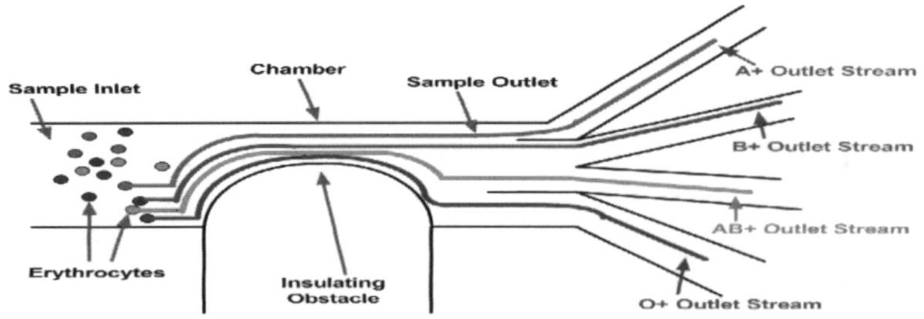

Figure 1: Insulating obstacle in a channel followed by channel bifurcations for sorting.

This attribute forms the foundation for the continuous flow, DC dielectrophoretic research outlined here. In DC-DEP, successful separation of cells into outlet channels is dependent on the deflection from the insulating obstacle (Figure 1). The preferred blood type channel is hypothesized from AC-DEP erythrocyte response results from previous research [3]. Cell accumulation at the outlet channel end ports will correspond to a specific blood type.

PDMS obstacles of varying geometrical configurations (rectangle, trapezium, and semi-circle) were utilized to create a spatially non-uniform electric field and blood types (A^+, B^+, AB^+, O^+, A^-, B^-, AB^-, and O^-) will be tested as they age (day 0, 1, 3 and 5). Results on fabrication and dependency studies: polystyrene experiments for size dependency, dilution experiments and optimizing the device design and selection of specific obstacle geometry have been presented here. Future goals are to fabricate microdevices with new optimized design parameters and again test for size dependency (fluorescent polystyrene particles) and antigen expression (red blood cell) dependency. This developed technique can be directly applied for use in portable blood diagnostic devices for easy, accurate and rapid analysis. Dielectrophoretic characterizations in lab-on-a-chip devices would be far more portable than current assay techniques.

Theory

In DC dielectrophoresis, an insulating obstacle is utilized to create a spatially non-uniform electric field. The movement of the particle in the electric field gradient is due to the resulting dielectrophoretic force acting on the polarizable particle [1, 2, 15]. The non-uniform field force can be derived from the net dielectric force as shown:

$$F = (p \cdot \nabla)E \qquad (1)$$

where 'p' is the dipole moment vector which is a function of particle's effective polarizability, 'α', volume, 'v', of the particle and the applied electric field, 'E' as shown:

$$p = \alpha v E \qquad (2)$$

The Claussius-Mossotti factor, α estimates the effective polarizability term of the particle and is a ratio of complex permittivities $\tilde{\varepsilon}$

$$\tilde{\varepsilon} = \varepsilon - i\sigma/\omega \qquad (3)$$

where 'ω' represents frequency, 'ε' the dielectric constant and 'σ' the electrical conductivity of the medium. There is no frequency component involved in DC-DEP and the dielectrophoretic force is estimated to be the residual of Claussius-Mossotti factor when the limit of the frequency approaches zero. This simplification is substituted into the dielectrophoretic field force equation:

$$F_{DEP} = \frac{1}{2} v \frac{\sigma_p - \sigma_m}{\sigma_p + 2\sigma_m} \nabla E^2 \qquad (4)$$

A spatially dense non-uniform field is created around the obstacle as the DC field lines diverge around the insulator. Due to the insulating obstacle, a high electric field density region is produced between the obstacle and the channel wall which aids in particle motion. Since DC-DEP forces push the particle towards or away from the high field density regions, the particle experiences an attractive or repulsive force as it flows around the corner of the obstacle, thus facilitating particle motion according to its polarizability. Many geometrical configurations and operating conditions are possible as a result of DC dielectrophoresis. In this work, a DC electric field is used to generate a field inside a channel, but around a rectangle, trapezium and semi-circle obstacle. Two dependencies are studied: size dependency using fluorescent polystyrene particles and antigen dependency using red blood cells of each blood type.

Materials and Methods

Variables in this research are: voltage of the DC source, device geometry, fluorescent polystyrene particles, blood type, and time of blood storage. Blood samples will be analyzed in the dielectrophoretic field on day 0 (day of donation), 1, 3 and 5 in storage at 5°C. The important steps in the process are device design, microdevice fabrication by soft lithography, experimentation (including microsample preparation), image analysis and quantification via total cells in each outlet channel. Each of these steps is discussed below:

Red Blood Cells (RBC) / Erythrocytes

They are biconcave in shape to maximize surface area for gas (O_2 and CO_2) transport across their membrane. Their diameters range between 6 to 8 microns with changes in response to solvent conditions, pH and temperature. RBC's have a net negative charge based on their conventional linear electrophoretic behavior [16] with their outer membrane being nonconductive ($\sigma \leq 1$ μS/m) [17] and the interior being conductive ($\sigma = 0.53$ to 0.31 S/m) [18]. Human blood types are classified based on (1) membrane surface antigens that span the entire membrane and (2) blood plasma antibodies [16]. Type A blood expresses antigen A, Type B blood expresses antigen B, Type AB blood has both A and B antigens while Type O blood has neither antigens [16]. The A and B antigen differ by side chain modifications. The A antigen terminates in an α 1,3 – linked N-acetylgalactosamine while B terminates in an α 1,3 – linked galactose. The presence or absence of the Rhesus (Rh) factor antigen determines the positive (present) and negative (absent) blood types [19]. The 8 ABO / Rh blood types are A^+, B^+, AB^+, O^+, A^-, B^-, AB^-, and O^-.

Device Design

AutoCAD software (Autodesk, Inc) was used to design the microdevice mask, which was printed on transparencies with a resolution of 32,512 dpi (Fine-Line Imaging, Inc.). The device consists of one inlet (200 μm wide) and 4 outlet channels (100 μm wide). The distance between the inlet and outlet ports was 4.5 cm. An insulating obstacle of varying geometry was positioned in the inlet channel at a distance of 150 μm upstream from the bifurcation to the four outlet channels. The obstacle geometries (rectangle, trapezium and semi-circle) are shown in Figure 2 (a), (b) and (c) with the complete microdevice design shown in Figure 2 (d).

(a) (b) (c)

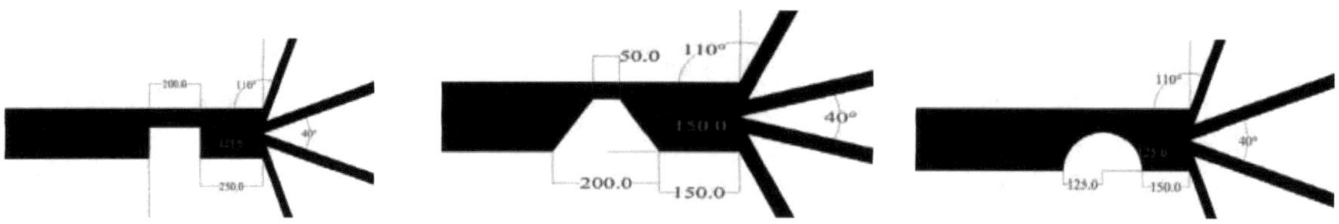

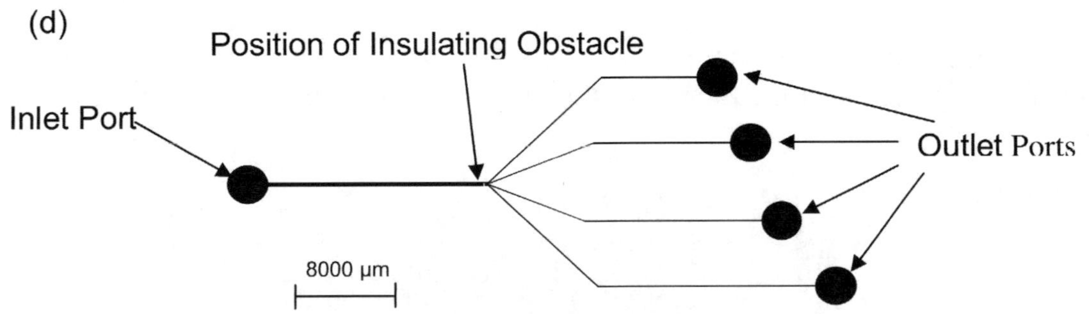

Figure 2: Insulating obstacle geometries and microdevice with inlet and outlet channels (a) Rectangular Obstacle Geometry (b) Trapezium Obstacle Geometry (c) Semi-Circle Obstacle Geometry (d) Microdevice design with inlet and outlet channel ports

Fabrication

The device was fabricated with a sample injection port and four outlet ports. This microdevice consisted of a glass slide and lithographically patterned microchannel of poly(dimethylsiloxane) (PDMS). Standard techniques of soft lithography were used to pattern the microdevice [12]. Some modifications in the process included employing UV-Ozone (UVO) cleaning procedure before SU-8 negative photoresist coating on the wafer, increasing the exposure time with the masked UV light and soft / hard bake temperatures and times. Figure 3 shows the patterned silicon wafer for different obstacle geometry. For the rectangular obstacle geometry, the inlet channel width of the patterned master (silicon wafer) increased by 20% compared to the mask dimensions, the width of the obstacle decreased by 11%, the distance between the obstacle and the channel wall increased by 13%, and the distance from the obstacle to the bifurcation point matched with that of the mask. For trapezium and semi-circle obstacle geometries, the width of inlet channel increased by 14%, width of obstacle decreased by 18%, distance between channel wall and the obstacle increased by 50% and the distance from obstacle to bifurcation point decreased by 22%.

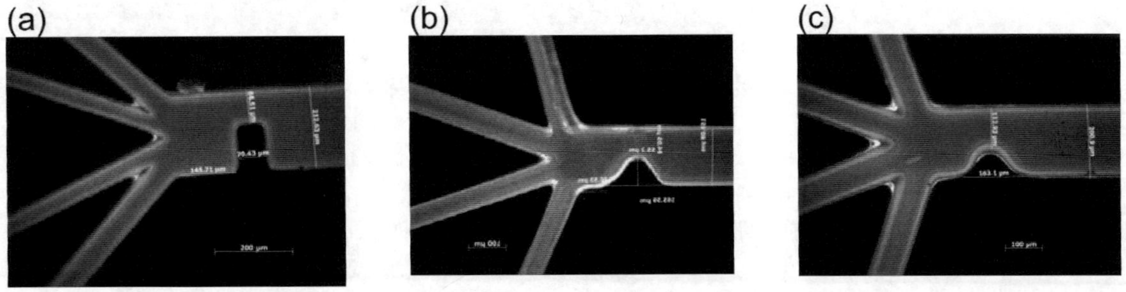

Figure 3: Patterned silicon wafer with rectangle, trapezium and semi-circle insulating obstacles. (a) Rectangular Obstacle Geometry (b) Trapezium Obstacle Geometry (c) Semi-Circle Obstacle Geometry.

PDMS was used to produce the microdevice from the patterned silicon wafer. Dow Chemical's Sylgard 184 pre-polymer base and curing agent were mixed in the ratio of 10:1, degassed for 15 minutes inside a vacuum chamber to remove suspended bubbles and poured onto the patterned negative relief silicon wafer. This wafer-PDMS assembly was cured in a 65°C oven for 12 hrs. The PDMS layer was peeled from the patterned silicon wafer and sealed onto a glass slide using UVO treatments. Biopsy punches were used to create inlet and outlet ports of 3 mm in diameters. Electrodes were positioned in the inlet and outlet ports and then connected to DC power supply. Figure 4 shows the PDMS microdevice on a glass slide with electrodes in the ports.

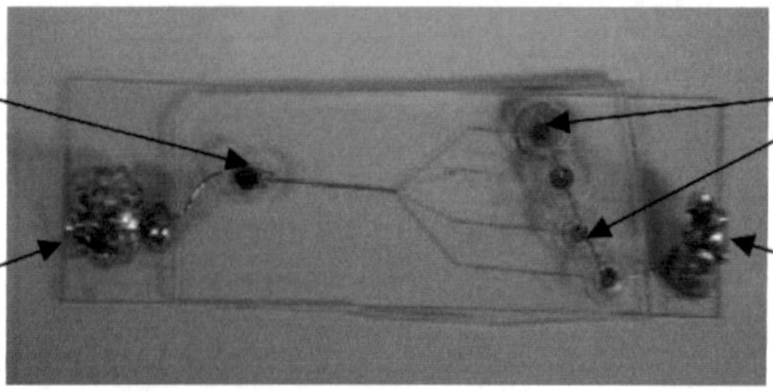

Figure 4: Typical PDMS microdevice sealed to a glass slide with electric connections and red food color in the channels for contrast.

Operation

DC dielectrophoretic force acting on a particle is proportional to its volume and subsequently size [5]. Therefore, two dependencies are simultaneously explored in this work: red blood cell type and blood cell size. This is possible because blood type antigens are expressed only on the red blood cells (erythrocytes) while whole blood is comprised of three different cells ranging in size from 2 μm (platelets), 7 μm (erythrocytes), and 11 μm (leukocytes). All work was conducted in a certified Biosafety Level II laboratory (BSL II) with Institutional Biosafety Committee (IBC) and Institute Review Board (IRB) for human subject's approval.

Polystyrene Particle Experiments

Fluorescent polystyrene particles (Bangs Laboratory, Inc) of three different diameters 2.28 μm (yellow, 540 or 600 nm emission), 5.49 μm (red, 660 or 690 nm emission) and 10.35 μm (green, 480 or 520 nm emission) were selected to roughly correspond to platelets, erythrocytes and leukocytes respectively. These particles were tested as a size dependence reference and will be compared against the separation and collection of actual blood cells into different channels in the microdevice shown in Figure 4. Based on the separation of these particles in all the three designs and the ease of fabrication, single obstacle geometry of rectangle was selected for further analyzing the blood samples. Individual as well as mixtures of all three sizes of polystyrene particles were used for experiments. Fluorescent polystyrene particles were diluted to 1:100 with 0.14M phosphate buffer saline (PBS) and introduced into the device through the inlet port for individual polystyrene experiments. A DC power supply was used to deliver a voltage of 5 V (field strength of 1.11 V_{pp}/cm) for separation of the particles over a distance of 4.5 cm between the inlet and outlet ports.

For mixture of the polystyrene particles experiments, a concentration of 1:100 was chosen. Still images of the inlet and outlet ports were captured by fluorescent microscopy at the start of the experiment and of the outlet ports at the end of the experiment and intensity changes noted. Total intensity change was tabulated after the experimental run (3 minutes) at each outlet ports to know the percent of particles travelled to each outlet port. The port having the maximum intensity change for specific wavelength corresponds to accumulation of a particular sized particle. Images were recorded using fluorescent microscopy every 2 seconds for 30 seconds. This interval time will aid in calculating the particle velocities and deflection from the obstacle, to infer particle trajectories to outlet ports. Based on the separation of these particles in the rectangular, trapezium, and semi-circle designs, the rectangular obstacle geometry was selected for further analysis of the blood samples.

Red Blood Cell Experiments

Whole blood was obtained through venipuncture by trained phlebotomists, drawn into vacutainers (Becton Dickinson) containing 1.8 mg K_2 EDTA per mL of blood, and stored at 4°C in a Biosafety Level 2 refrigerator. 0.25 ml of whole blood sample was taken in a centrifuge vial and diluted to comprise 2 ml of sample volume with 0.14M of phosphate buffer saline (PBS) (0.1 S/m) with initial weight of the sample mixture being recorded. The sample with PBS was mixed thoroughly and centrifuged at 2000 rpm for 5 minutes. The plasma/PBS layer was decanted off and red blood cells were again resuspended in PBS to the same initial weight, mixed and recentrifuged. This process of centrifuging and washing with PBS was conducted three times and the red blood cell mixture at the end of the washing procedure was used for further experiments. This washing procedure was conducted to isolate red blood cell antigen responses in the field from antibody interactions from the plasma. To determine the appropriate dilution of the blood samples, systematic dilution experiments were conducted. Centrifuged red blood cells were diluted using PBS to obtain various concentration levels of 1:60, 1:100, 1:600, and 1:800. Each dilution was loaded into the microdevice separately and imaged using light microscopy. In this experiment for DC dielectrophoresis, the number of red blood cells/ unit area in the field of view is matched to prior data counts of $2.5*10^{-8}$ cells/ μm^2.

Centrifuged red blood cell sample of 1:600 concentration from the dilution experiments were introduced via inlet port to the PDMS microdevice (shown in Figure 4) mounted on Zeiss fluorescent microscope. A DC power supply was used to deliver a range of voltages (10 V-200 V) to determine optimal response of cells. High-resolution video microscopy was used to record bright field video of cell motion and outlet port cell counts within the microdevice.

To determine the total time and the intervals for image collection, systematic experiments will be conducted. The entire experiment will be conducted for total of 15 minutes with each experimental cycle of 3 minutes and 2 second intervals for image collection within the experimental cycle. A total of 3 experimental cycles will be conducted with a gap time of 3 minutes between each experimental cycle. Red blood cell responses to the electric field are slow compared to polystyrene particles and faster imaging speed is not required. Cell movement will be recorded via light microscopy between the insulating obstacle and the bifurcation point. These raw images from the microscope will be used for analysis of velocities of the cells and subsequent travel to the outlet channel. Once the time of experimental run and the interval for image collection is determined, this will be used for further experiments. Cell counts will be tallied in each outlet port at the completion of each run to assess the hypothesis that the port having the maximum cells likely corresponds to a specific blood type. Experiments are run over day 0 (day of collection), 1, 3 and 5 of blood storage for each blood type [16].

Results and Discussions

The quantitative and qualitative analysis of blood types will be demonstrated via DC dielectrophoresis in this work. Here two dependencies will be studied: size dependency via polystyrene experiments and antigen dependency via red blood cell experiments.

Polystyrene Particle Experiments

From size dependency experiments conducted using pure fluorescent polystyrene particles on rectangular obstacle geometry, the authors observed that the 10.35 μm particles deflected closely around the obstacle and preferentially travelled to the lowermost channel (hypothesized O^+ outlet stream in Figure 1) as shown in Figure 5(a) whereas the 5.49 μm particles were not deflected much and travelled into the topmost channel (hypothesized A^+ outlet stream in Figure 1) as shown in Figure 5(b).

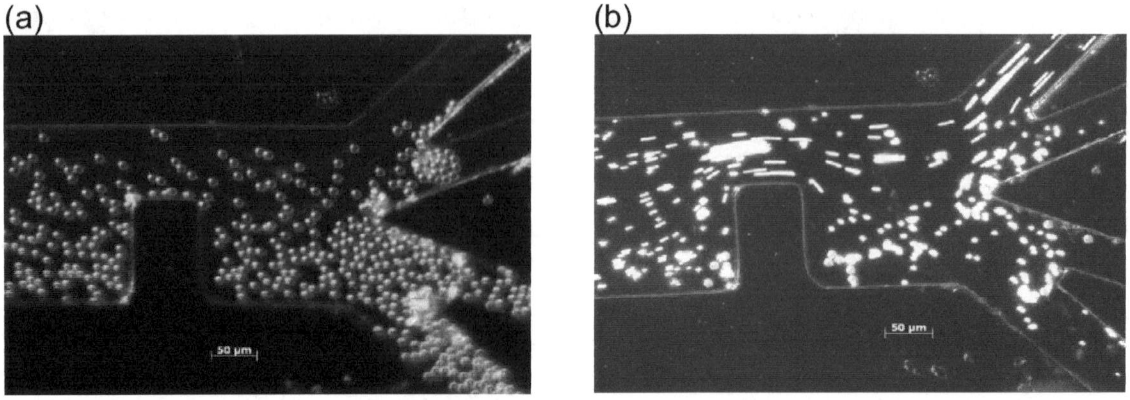

Figure 5: 10.35 μm and 5.49 μm particle trajectories in the rectangular obstacle channel. (a) 10.35 μm particles preferentially flow to channel # 4. (b) 5.49 μm particles preferentially flow to channel # 1.

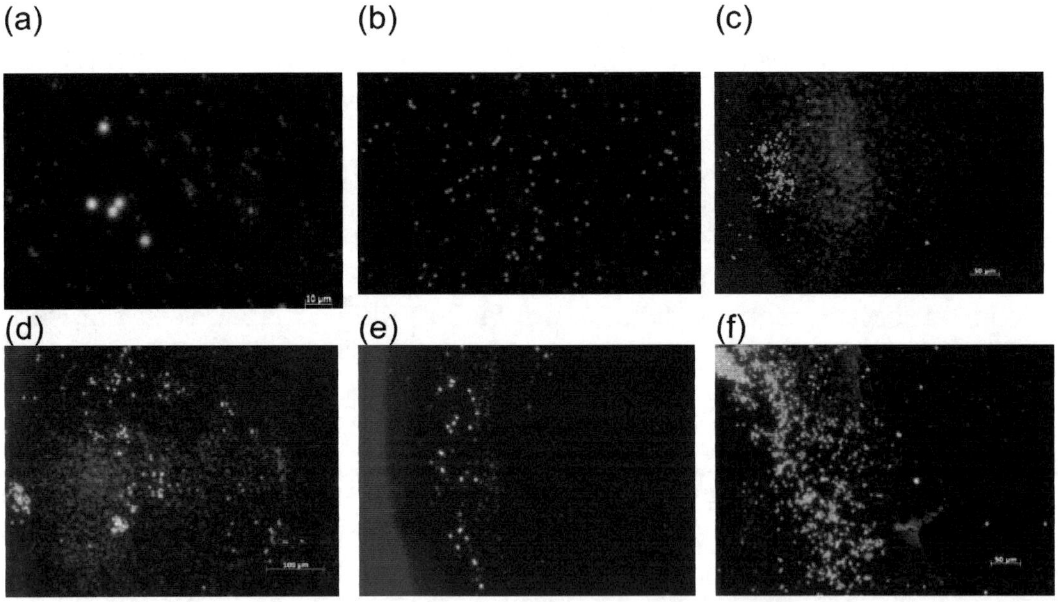

Figure 6: Fluorescent polystyrene particles imaged at the start and end of experimental runs at each ports. (a) Model representing different sizes of polystyrene particles: blue particles are 2.28 μm, dragon red particles are 5.49 μm and green particles are 10.35 μm. (b) Inlet port having all the three sizes of polystyrene particles at the start of the experiment. (c), (d), (e) and (f) Outlet ports 1, 2, 3 and 4 at the end of 3 minute experimental run.

Experiments were also conducted by mixing all the three different diameter polystyrene particles of concentration 1:100 by using 0.14M PBS and are shown in Figure 6 to analyze size dependency. Continuous images were taken for approximately 3 minutes with an interval of 2 seconds. Intensity change will be measured for all the four outlet ports and the port having maximum intensity change for a specific wavelength corresponds to accumulation of a particular polystyrene particle size. The raw images from the microscope will be further analyzed to obtain particle velocities, as an additional indicator to the particle trajectory and subsequent motion into the outlet channels. Figure 6 shows the results of 3 minute experimental run of polystyrene particles. Outlet 1, which is the channel farthest from the base of the rectangular obstacle, has a high concentration of 2.28 μm and 5.49 μm particles (Figure 6(c)) whereas outlet 4 has a high concentration of 10.35 μm particles (Figure 6(f)). Outlet 2 has a high concentration of 5.49 μm particles (Figure 6(d)). This shows that 5.49 μm particles were the least deflected form the insulating obstacle and 10.35 μm particles were the highest deflected which matches with the individual experiment results shown in Figure 5 (a) and (b). Figure 7(a) shows the deflections of the mixture of polystyrene particles from the insulating obstacle into outlet channels. The separation of these particles can be improved by reducing the distance between the electrodes thus creating greater field strength at the obstacle.

Red Blood Cell Experiments

From previous AC-DEP studies by the authors, type A had the largest deflection from high field intensity region whereas type O had the least deflection in the nonuniform electric field. These separations of blood types were achieved in 60 seconds [3]. To achieve a comparable separation in DC-DEP, an intense dielectrophoretic field is necessary. This deflection pattern has been attributed to the existence of antigens on the membrane surface. On this basis, authors are predicting the same pattern of deflection to be followed in DC dielectrophoresis experiments while the blood types separate at the bifurcation point into channels. Images from the microscope will be further analyzed to obtain cell movement. Total cell counts will be tabulated after the entire experimental run to know the percent of cells travelled to each outlet port.

(a) (b)

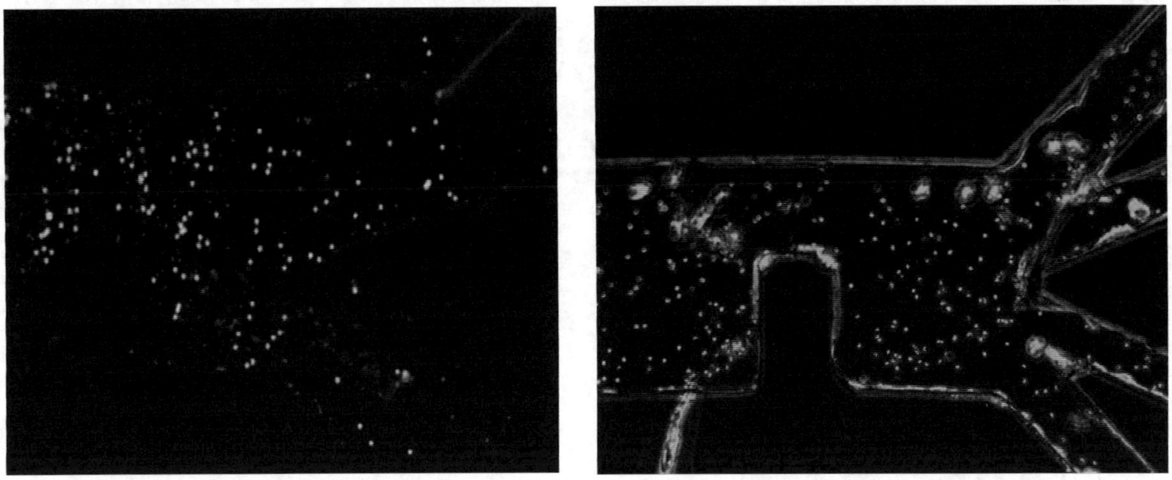

Figure 7: Deflection from the insulating obstacle and particle trajectories for (a) polystyrene particles, and (b) red blood cells.

O^- blood type was introduced into the rectangular geometry-based channel. The authors observed during this experiment that electric field density between the electrodes was not sufficient for separation to occur (Figure 7(b)). The voltage applied through a DC source was 190 V which resulted in electric field strength of 0.00422 V_{pp}/ µm due to the longer distance of 4.5 cm between the electrodes. However, a future design will aid in improving and optimizing device parameters for future experiments; the distance between the inlet and outlet ports (i.e. electrode spacing) has been reduced to 1 cm from 4.5 cm, the distance between channel wall and the obstacle has been reduced to 15 µm, 30 µm and 50 µm from 75 µm and the distance between the insulating obstacle and bifurcation point has been reduced to 50 µm, 75 µm and 100 µm from 250 µm. Thus by optimizing the design parameters we can achieve greater field strength between the obstacle and channel wall thus aiding in reduced separation times compared to AC dielectrophoresis.

Summary

The use of lab-on-a-chip devices for blood typing is a novel technique which will potentially aid in increasing the speed and efficiency of blood type diagnosis. DC dielectrophoresis shows excellent potential for this purpose. Here we have demonstrated separation of erythrocyte responses and polystyrene particles via insulator-based dielectrophoresis. Fabrication of these microdevices was achieved by soft photolithography and PDMS casting. Three geometries (rectangle, semi-circle and trapezium) were assessed for DC dielectrophoretic movement, but the rectangle geometry was selected for further experiments due to better separations achieved during polystyrene particle experiments. From previous research on AC-DEP, four parameters were used to quantify the cell movement based on their blood types: total cell count, vertical movement, horizontal movement and combined distance. The best separation was achieved in 60 seconds with type O being the least deflected and type A being the most deflected from high field density regions [3]. Future work will be to optimize the geometry of the DC-DEP microdevice to achieve similar blood type separations. By optimizing the geometry of the microdevice, the DC voltage applied between the electrodes can be decreased thus increasing the electric field strength between the channel wall and insulating obstacle to maximize electric field density gradients. Increases in electric field density are expected to yield much better separations of blood type erythrocytes. Currently with the rectangular geometry, we have separated 2.28 µm, 5.49 µm and 10.35 µm polystyrene particles at 5 V / 4.5 cm and also have conducted experiments on type O^- red blood cell for optimizing geometry of the device.

References

1. Minerick, A. R. (2008), "DC Dielectrophoresis in Lab-on-a-Chip Devices." In: Li, Dongqing (ed). *Encyclopedia of Micro- & Nanofluidics*. Springer, Berlin Heidelberg New York (in press).
2. Keshavamurthy S.S., Leonard K.M., Burgess S.C., Minerick A.R. (2008), "Direct current dielectrophoretic characterization of erythrocytes: Positive ABO blood types", *NSTI-nanotech*, 3, 401–404.

3. Srivastava S.K., Daggolu P.R., Burgess S.C., Minerick A.R. (2008), "Dielectrophoretic characterization of erythrocytes: Positive ABO blood types", *Electrophoresis*, in press.
4. Minerick, A.R. (2008), "The Rapidly Growing Field of Micro and Nanotechnology to Measure Living Cells," Invited Perspective with Cover Figure, *AIChE Journal*, 54(9), 2230-2237.
5. Lapizco-Encinas B.H., Simmons B.A., Cummings E.B., Fintschenko Y. (2004), "Dielectrophoretic Concentration and Separation of Live and Dead Bacteria in an Array of Insulators", *Analytical chemistry*, 76, 1571-1579.
6. Kang K, Kang Y, Xuan X, Li D. (2006), "Continuous separation of microparticles by size with Direct current-dielectrophoresis", *Electrophoresis*, 27, 694-702.
7. Cummings E.B., Singh A.K. (2003), "Dielectrophoresis in microchips containing arrays of insulating posts: Theoretical and experimental results", *Analytical chemistry*, 75, 4724-4731.
8. Mela P., Van der berg A., Fintschenko Y., Cummings E.B., Kirby B.J. (2005), "The zeta potential of cyclo-olefin polymer microchannels and its effects on insulative (electrodeless) dielectrophoresis particle trapping devices", *Electrophoresis*, 26, 1792-1799.
9. Barbulovic-Nad I., Xuan X., Lee J.S.H., Li D. (2006), "DC-dielectrophoretic separation of microparticles using an oil droplet obstacle", *Lab on a chip*, 6, 274-279.
10. Thwar P.K., Linderman J.J., Burns M.A. (2007), "Electrodeless direct current dielectrophoresis using reconfigurable field-shaping oil barriers", *Electrophoresis*, 28, 4572-4581.
11. Zhang, L.; Tatar, F.; Turmezei, P.; Bastemeijer, J.; Mollinger, J. R.; Piciu, O.; Bossche, A. (2006), "Continuous Electrodeless Dielectrophoretic Separation in a Circular Channel", *Journal of Physics*, 34, 527-532.
12. Anderson J.R., Chiu D.T., Jackman R.J., Chernlavskaya O., McDonald J.C., Wu H.K., Whitesides S.H., Whitesides G.M. (2000), "Fabrication of topologically complex three-dimensional microfluidic systems in PDMS by rapid prototyping", *Analytical Chemistry*, 72(14), 3158–3164.
13. Kang, Y.; Li, D.; Kalams, S. A.; Eid, J. E. (2008), "DC-Dielectrophoretic separation of biological cells by size", *Biomed Microdevices*, 10(2), 243-249.
14. Pysher, M. D., Hayes, M. A. (2007), "Electrophoretic and Dielectrophoretic Field Gradient Technique for Separating Bioparticles", *Analytical Chemistry*, 79 (12), 4552-4557.
15. Pohl, H. (1978), "*Dielectrophoresis: The behavior of neutral matter in nonuniform electric fields*", New York, NY, Cambridge University Press.
16. Patenaude S.I., Sato N.O.L., Borisova S.N., Szpacenko A. (2002), "The structural basis for specificity in human ABO(H) blood group biosynthesis", *Nature structural biology*, 9, 685 – 690.
17. Gasocyne P., Mahidol C., Ruchirawat M., Satayavivad J., Watcharasit P., Becker F.F. (2002), "Microsample preparation by dielectrophoresis: isolation of malaria", *Lab on a Chip*, 2, 70-75.
18. Gimsa J., Muller T., Schnelle T., Fuhr G. (1996), "Dielectric spectroscopy of single human erythrocytes at physiological ionic strength: dispersion of the cytoplasm", *Biophys. J.*, 71, 495-506.
19. Daniels G., Bromilow I. 92007), *"Essentials guide to Blood Groups"*, Blackwell Publishing.

AUTHOR INDEX

AUTHOR INDEX

AUTHOR INDEX